ARE CHEMICAL JOURNALS TOO EXPENSIVE AND INACCESSIBLE?

A WORKSHOP SUMMARY TO THE CHEMICAL SCIENCES ROUNDTABLE

Ned D. Heindel, Tina M. Masciangioli, and Eva von Schaper, Editors

Chemical Sciences Roundtable

Board on Chemical Sciences and Technology

Division on Earth and Life Studies

NATIONAL RESEARCH COUNCIL
OF THE NATIONAL ACADEMIES

THE NATIONAL ACADEMIES PRESS
Washington, D.C.
www.nap.edu

THE NATIONAL ACADEMIES PRESS 500 Fifth Street, N.W. Washington, DC 20001

NOTICE: The project that is the subject of this report was approved by the Governing Board of the National Research Council, whose members are drawn from the councils of the National Academy of Sciences, the National Academy of Engineering, and the Institute of Medicine. The members of the organizing committee responsible for the workshop summary were chosen for their special competences and with regard for appropriate balance.

This study was supported by Research Corporation under Grant GG0066, the U.S. Department of Energy under Grant DE-AT01-04ER15535, the National Institutes of Health under Grant N01-OD-4-2139 (Task Order 25), and the National Science Foundation under Grant CHE-0328197.

Any opinions, findings, conclusions, or recommendations expressed in this publication are those of the authors and do not necessarily reflect the views of the organizations or agencies that provided support for the project.

International Standard Book Number 0-309-09590-5

Additional copies of this report are available from the National Academies Press, 500 Fifth Street, N.W. Lockbox 285, Washington, DC 20055; (800) 624-6242 or (202) 334-3313 (in the Washington metropolitan area); Internet, http://www.nap.edu

Printed in the United States of America

CHEMICAL SCIENCES ROUNDTABLE

Co-chairs

F. FLEMING CRIM (NAS), University of Wisconsin, Madison, WI
MARY L. MANDICH, Lucent Technologies, Murray Hill, NJ

Members

PAUL T. ANASTAS, Green Chemistry Institute, Washington, DC
PATRICIA A. BAISDEN, Lawrence Livermore National Laboratory, Livermore, CA
MICHAEL R. BERMAN, Air Force Office of Scientific Research, Arlington, VA
RICHARD BUCKIUS, National Science Foundation, Arlington, VA
LEONARD J. BUCKLEY, Defense Advanced Research Projects Agency, Arlington, VA
CHARLES P. CASEY (NAS), University of Wisconsin, Madison, WI
ARTHUR B. ELLIS, National Science Foundation, Arlington, VA
TERESA FRYBERGER, Office of Science and Technology Policy, Washington, DC
JEAN H. FUTRELL, Pacific Northwest National Laboratory, Richland, WA
PAUL GILMAN, Environmental Protection Agency, Washington, DC
ALEX HARRIS, Brookhaven National Laboratory, Upton, NY
SHARON HAYNIE, E. I. du Pont de Nemours & Company, Wilmington, DE
NED D. HEINDEL, Lehigh University, Bethlehem, PA
CAROL J. HENRY, American Chemistry Council, Arlington, VA
MICHAEL J. HOLLAND, Office of Science and Technology Policy, Washington, DC
CHARLES T. KRESGE, Dow Chemical Company, Midland, MI
GEORGE H. LORIMER (NAS), University of Maryland, College Park, MD
PAUL F. MCKENZIE, Bristol-Myers Squibb Company, New Brunswick, NJ
WILLIAM S. REES, Department of Homeland Security, Washington, DC
GERALDINE L. RICHMOND, University of Oregon, Eugene, OR
MICHAEL E. ROGERS, National Institutes of Health, Bethesda, MD
JEFFREY J. SIIROLA (NAE), Eastman Chemical Company, Kingsport, TN
DOTSEVI Y. SOGAH, Cornell University, Ithaca, NY
WALTER J. STEVENS, Department of Energy, Washington, DC
FRANKIE WOOD-BLACK, ConocoPhillips, Houston, TX

National Research Council Staff

KAREN LAI, Research Associate, Christine Mirzayan Graduate Fellow (August-December 2004)
TINA M. MASCIANGIOLI, Program Officer
ERICKA M. MCGOWAN, Research Associate
DAVID C. RASMUSSEN, Project Assistant
DOROTHY ZOLANDZ, Director

BOARD ON CHEMICAL SCIENCES AND TECHNOLOGY

Preface

The Chemical Sciences Roundtable (CSR) was established in 1997 by the National Research Council. It provides a science-oriented apolitical forum for leaders in the chemical sciences to discuss chemistry-related issues affecting government, industry, and universities. Organized by the National Research Council's Board on Chemical Sciences and Technology, the CSR aims to strengthen the chemical sciences by fostering communication among the people and organizations—spanning industry, government, universities, and professional associations—involved with the chemical enterprise. One way it does this is by organizing workshops that address issues in chemical science and technology that require national attention.

In October 2004, the CSR organized a workshop on the topic, "Are Chemical Journals Too Expensive and Inaccessible?" This workshop provided a forum to discuss the publication of chemistry journals within the larger context of scientific, technical, and medical (STM) journal publishing. Issues relevant to the different stakeholders from academe, industry, and government were addressed, such as whether the needs of users of chemical information are being met; how librarians are responding to changes in STM publishing; the economics of publishing chemical journals; and whether the increasing cost of subscriptions is presenting obstacles to carrying out research in chemistry and chemical engineering. As part of this activity, the unique scientific journal needs of chemists and chemical engineers and the new approaches for addressing those needs—including "open access"—were explored.

This document summarizes the presentations and discussions that took place at the workshop, which have been edited and organized around the major themes of historical perspective, challenges of web publication, cost, access, archives, and open access. In accordance with the policies of the CSR, the workshop *did not* attempt to establish any conclusions or recommendations about needs and future directions, focusing instead on issues identified by the speakers.

Ned D. Heindel
Workshop Organizer

Acknowledgment of Reviewers

This workshop summary has been reviewed in draft form by persons chosen for their diverse perspectives and technical expertise in accordance with procedures approved by the National Research Council's Report Review Committee. The purpose of this independent review is to provide candid and critical comments that will assist the institution in making its published workshop summary as sound as possible and to ensure that the summary meets institutional standards of objectivity, evidence, and responsiveness to the workshop charge. The review comments and draft manuscript remain confidential to protect the integrity of the deliberative process. We wish to thank the following individuals for their review of this workshop summary:

Carol Tenopir, University of Tennessee
Philip Barnett, City College of New York
Gordon Tibbitts, Blackwell Publishing Inc. (US)
Marc H. Brodsky, American Institute of Physics
Claude F. Meares, University of California-Davis
Carol Carr, University of Pennsylvania
Jack Halpern, University of Chicago

Although the reviewers listed above have provided many constructive comments and suggestions, they did not see the final draft of the workshop summary before its release. The review of this workshop summary was overseen by **C. Herb Ward**, of Rice University. Appointed by the Division on Earth and Life Studies, he was responsible for making certain that an independent examination of this workshop summary was carried out in accordance with institutional procedures and that all review comments were carefully considered. Responsibility for the final content of this workshop summary rests entirely with the authors and the institution.

Contents

1

Overview

This day and a half workshop began with a historical look at communicating science and the origins of chemical journals. **Arnold Thackray**, president of the Chemical Heritage Foundation, took participants back 350 years to the establishment of the first scientific publication—*Philosophical Transactions*—when scientists first recognized the need to establish priority of scientific discovery. The picture he painted of the development of the chemical journal from seventeenth century is strikingly similar to the current state of scientific publishing—overwhelming quantities and varying quality of data, high publishing costs, rapidly advancing information technology, and the emergence of new scientific disciplines and subdisciplines of chemistry.

The remainder of the workshop explored in more detail the unique ways in which chemists use the scientific literature, whether they have the access to the quantity and quality of journals they need, and the new approaches being taken to make journals more accessible and of highest impact.

UNIQUE JOURNAL NEEDS OF CHEMISTS AND CHEMICAL ENGINEERS

The unique scientific journal needs of chemists and chemical engineers were discussed from the perspective of a chemistry society publisher (**Robert Bovenschulte**, American Chemical Society), an academic chemist (**Christopher Reed**, University of California, Riverside), a commercial chemical journal publisher (**Patrick Jackson**, Elsevier), a university chemistry librarian (**Andrea Twiss-Brooks**, University of Chicago), and an ACS journal editor (**Gordon Hammes**, University of North Carolina, Chapel Hill), and are briefly described below:

The speakers described chemistry as a core discipline, where material often has to be made easily accessible to non-chemistry experts, but at the same time provide detailed content for practicing chemists. Chemistry is also a very visual discipline with its own language. Other unique needs of chemists discussed include their special graphical needs, such as drawing tools for chemical structures, reactions, and other illustrations. They also need visualizing tools to search the literature, such as graphical tables of contents that provide an associative representation of document contents that can be chemical structures or reactions in a journal article. There was discussion about chemists and biochemists relying more on their indexing and abstracting services than some other scientists.

In terms of searching the literature, chemists were described as being self-reliant and self-instructed. Despite all of the advances in technology, browsing is still an important means of finding literature., and scanning journals continues to be handed down from the faculty adviser to graduate students. Databases were presented as important tools for chemists and chemical engineers, but it was pointed out that users tend to rely heavily on one or two databases with which they are familiar. At the same time, it was mentioned that chemists do not entirely trust tools like Current-Contents or table-of-contents alerting systems; they fear that these are not dynamic enough to keep up with their changing interests and that they might miss information. Therefore, they will also search the literature manually.

ACCESS TO, AND IMPACT OF, CHEMISTRY JOURNALS

Gaining access to the full quantity and quality of chemical journals was addressed in the next workshop session—from the perspective of an open-access interactive peer-reviewed journal editor (**Ulrich Pöschl**, Technical University of Munich and *Atmospheric Chemistry and Physics*), an industrial librarian (**Lou Ann DiNallo**, Bristol Myers-Squibb), an academic chemist working with mainly undergraduate students (**Michael Doyle**, University of Maryland), an academic chemist (**R. Stephan Berry**, University of Chicago), and a chemistry society publisher (**Peter Gregory**, Royal Society of Chemistry). In this discussion, the speakers largely addressed the access issue by examining the overall journal publication process.

Speakers and participants expressed their desire for rapid processing of journal article submissions, which has been significantly improved by electronic capabilities. However, many felt that there is still no way to overcome the slowness of the peer-review process, and keeping up the quality of a publication through rigorous peer review is viewed as essential to the credibility of journals.

There was significant discussion about the implications of publishing in high-profile journals and influence of journal impact on where authors publish. High profile (high-impact) journals such as *Science* and *Nature* were criticized for their low acceptance rates, which some workshop participants feel leads to some important research being overlooked. Another criticism was made of how many papers in these journals end up being too brief with little or no data to refer to. It is often difficult to know how the experiments were done and whether they are reproducible in individual labs.

Participants discussed how journal impact and quality is determined—especially the heavy use of what is known as the impact factor. According to Thompson ISI (Institute for Scientific Information), the impact factor is a measure of the frequency with which an average article, published within the last two years in a journal, has been cited in a particular year or period. Some thought that the impact factor was highly overrated and that it should be looked at, but not very seriously. It was pointed out that there is a tremendous difference between various subfields in terms of impact factor, with well-established life sciences journals having far higher impact factors than chemistry journals and applied journals. However, in libraries, which face cutting journal subscriptions, impact factor is not the only thing that determines which journal to cut because there are journals that are very heavily read but not frequently cited.

There was also concern from the workshop participants about the proliferation of journals, which is thought as one cause of the rising cost of accessing journals. In new fields such as proteomics new journals are needed, but in most cases some feel that the new journals are just duplicating other journals. However, one participant cautioned that a decline in the number of journals might make it difficult to publish really adventurous work.

Access to archives was discussed with much interest. Some workshop attendees consider many chemical journals to be at the bottom of their class in regard to access to archives, as well as author rights. Authors in chemical journals often do not retain the right to use their work in subsequent compilations, post it on their web sites, post and update it on e-print service, or automatically grant third-party noncommercial use. A number of participants expressed their dissatisfaction with the slowness of the American Chemical Society to improve this situation.

Rising library costs were a major concern of some workshop participants. Some feel that there is a concentration of scholarly output in the hands of a small but highly influential number of commercial publishers. This is considered to be leading to a disproportionate and rising level of library budgets being spent on journal subscriptions.

This session and the first day of the workshop closed with a talk by academic researcher and open-access advocate **Stevan Harnad**, (University of Quebec, Montreal) who discussed the different paths to maximizing research access and impact. He discussed the origins of the open-access movement and how open-access approaches lead to increased research impact.

NEW APPROACHES TO ADDRESS JOURNAL NEEDS

In this third and final workshop session, participants considered ways to improve communication of science through making journals more accessible. Presentations were given from the perspective of a society publisher experimenting with open-access (**Bridget C. Coughlin**, *Proceedings of the National Academy of Science*s (PNAS) and **Nicholas Cozzarelli**, University of California-Berkeley and *PNAS*), a representative from a scientific society (**Martin Apple**, Council of Science Society President), a university press publisher (**Michael Keller**, Stanford University Press), a scientific society publisher (**Martin Blume**, American Physical Society), a representative of an open-access journal (**Vivian Siegel**, Public Library of Science-PLoS), and a university library database manager (**Anna Gold**, Massachusetts Institute of Technology).

Various publishing and associated financing models were presented—subscription based, pay-per-view, and author pays. There seemed to be agreement among many participants that open access—where articles are freely available to anyone on-line immediately upon publication—is a good idea, but the funding mechanism for implementing it is not as clear. One of the current open-access models discussed extensively was the author-pays—where the author rather than subscriber pays for publishing services and the final paper is published in an open access journal. However, many participants felt that a major issue with the author-pays model is that in the existing culture of chemistry, authors do not pay page charges. According to some workshop participants, the open-access experiments of *PNAS*, PLoS, and *Atmospheric Chemistry and Physics* may not be suited to most chemical journals, so other approaches are needed.

It was pointed out that older literature is more heavily utilized in chemistry than is the case for other disciplines. Short of making all current research freely available online, free access to journal back-files, self-archiving, and institutional repositories were presented as other approaches to making more research accessible. In terms of self-archiving a manuscript or a document, chemists and other scientists apparently are adverse to do this, and chemical publisher copyright rules do not currently support self-archiving approaches. Many participants urged chemical journal publishers to make their archives freely available online for at least a trial period. However, publishers explained how they had made large investments in digitizing and storing back-files, which prevents them from making archives freely available at this time.

Some participants expressed concern that if chemists, librarians, and publishers cannot make progress on these issues on their own, they will likely get a push from the federal government. At the time of the workshop there was a lot of discussion about legislation involving the National Institutes of Health, whereby NIH was proposing to require researchers to submit an electronic version of any publication that results from research supported, in whole or in part, with direct costs from NIH to the NIH National Library of Medicine's (NLM) PubMed Central (PMC)—a digital repository of full-text, peer-reviewed biomedical, behavioral, and clinical research journals. PMC is a publicly accessible, permanent, and searchable electronic archive available on the Internet at *http://www.pubmedcentral.nih.gov/*. Since the workshop took place, this NIH Public Access Policy—intended to enhance public access to archived publications resulting from NIH-funded research—has moved forward.[1] However, the language of the legislation has been softened after public comment from publishers, patient advocates, and scientists. Beginning May 2, 2005, authors are requested (not required) to submit an electronic version of published NIH-funded work to PMC. Additional information about the NIH Public Access Policy is available on the Internet at *http://www.nih.gov/about/publicaccess/index.htm*.[2]

[1]"Policy on Enhancing Public Access to Archived Publications Resulting From NIH-Funded Research," *Federal Register* 70(26): 6891-6900 (Federal Register Document 05–2542), (2005).

[2]Readers should also note that partially as a result of the public access guidelines recently released by the NIH, the American Chemical Society announced in March 2005 that they will allow free access to the full-text version of all research articles published in ACS journals via an author-directed Web link twelve months after final publication. This expands on the organization's current practice of permitting 50 downloads of authors' articles free of charge during the first year of publication. For more information see S. Rovner, "ACS Broadens Article Access—Conditions set for free availability one year after publication," *Chemical and Engineering News*, 83(10), 10, (2005).

2

Historical Perspective

Arnold Thackray illustrated the history of chemical journals, proposing to understand what is happening in the present by looking into the past.[1] "In every age, the amount of literature has seemed as if it is going to overwhelm people," Thackray said. Science has been about communication for the past 350 years, he continued. Both the science itself and the means of its communication are competitive enterprises, where the prize is to be first, he said. He talked about the origins of chemical journals, the rise of scientific societies, the idea of a "world brain" and journals after World War I.

THE ORIGINS OF CHEMICAL JOURNALS

Much of what is true today was true 350 years ago, when scientific journals originated. "The issue of how you pay for what you want to do was right there on the first day," Thackray said. At that time Robert Boyle, son of one of the richest men in England, funded the newly established Royal Society. One mission of the early Royal Society was that it would help to establish priority of discovery. "It finally got enough people poking about interrogating nature that the concern of, 'I was first,' and 'I thought of it,' was right there," Thackray said.

When one of Robert Doyle's employees, Henry Oldenburg, founded the first scientific journal, the problem of making it profitable soon arose. Oldenburg founded *Philosophical Transactions* (changed to *Philosophical Collections* by Robert Hook in 1677) in 1665 as a private venture, thinking he could make a profit from subscriptions paid to the Royal Society, but he never made more than 40 pounds a year, which was what it cost to rent his house.

Meanwhile, in Paris, at the same time, another independent journal was started up, *Le Journal des Sçavans (Journal of the Learned)*. It predated the founding of *Academie Royale des Sciences* in Paris, but no sooner had the journal started than there was a scientific gathering on which it could report. Members of the academy could buy personal subscriptions to this journal. The French Academy did not begin publishing its own journal, the *Histoires et Memoirs,* until 1699.

From the concept of establishing who was first in science and in the ensuing communication, it took about 100 years to develop what we today would call peer review. By 1752, the Royal Society had put a committee in charge of the selection of the papers, provoked by a book published in 1751, *A Review of the Works of the Royal Society of London, Containing Animadversions on Such of the Papers as Desire Particular Observations*. "Some guy called John Hill lampooned the quality of what was in *Phil. Trans.*," Thackray said.

The eighteenth century also brought the proliferation of journals, because entrepreneurs, many of whom were natural philosophers themselves, saw an opportunity for publishing. One of the most memorable journals, *Observations sur la Physique, sur* L'*Histoire Naturelle, et sur les Artes*, was published by Rozier and Mongez in 1778. At this point in history, "The number of scholars has increased. These motives led to the desire for a periodical supplied quickly and regularly which would announce the discoveries which are made each day in the different branches of the sciences. You can see that the pace is really picking up," Thackray said.

One of the more rapid communications in Rozier's journal was Antoine Lavoisier's paper on the burning of diamond. It illustrates the games you could play even then with publication, Thackray said. The paper that Lavoisier

[1]For more in-depth information on the history of chemistry and chemical journals presented here, visit the Chemical Heritage Foundation web site at *www.chemheritage.org*.

published in Rozier's journal in 1772 was published officially in the memoirs of the academy four years later, allowing him to profit from Joseph Priestley's work, which he learned of in the interim.

Soon after, the first journals that dealt solely with chemistry appeared. The *Chemisches Journal,* in 1778, by Lorenz von Crell was founded to encourage German chemistry, followed by the *Annales de Chimie*, in 1789 in France. The *Annales de Chimie* is the first journal to promote a program, since it only published chemistry in the then-new nomenclature. As late as 1800 Joseph Priestley published his *Doctrine of Phlogiston Established* in the United States, but from 1789 on, there was no longer any reference to phlogiston in the *Annales de Chimie*. Only things cast in Lavoisier's new theoretical scheme and nomenclature appeared there.

The high price of journals was also an issue back then. A volume could cost 3.5 pounds for 300 pages. Translated into how many days the average person would have to work to publish it, this is probably equivalent to about $1,000 today, Thackray said.

Then as now, new work appeared in new journals. The *Philosophical Transactions* of the Royal Society made no mention of electrochemistry. The new work in electrochemistry, by Humphry Davy, appeared in the *Journal of Natural Philosophy, Chemistry, and the Arts*, founded by the entrepreneur William Nicholson.

As is also true today, reading a journal did not necessarily mean being current in the newest science. Dalton's atomic theory, for example, was first published in a university textbook, written by Thomas Thompson. Thompson, who published the *Annals of Philosophy* in 1814, was very influential. He lectured to university students, and his *System of Chemistry* was the standard text. The third edition reads: "We have no direct means of ascertaining the density of the atoms of bodies, but Mr. Dalton, to whose uncommon ingenuity and sagacity the philosophic world is no stranger, has lately contrived an hypothesis which, if it proved correct, will furnish us with a very simple method of ascertaining that density with great precision." Thompson goes on to articulate Dalton's chemical atomic theory. Of course, the reason scientific research moved so slowly, Thackray added, was that the professional scientist had not yet appeared. "Dalton didn't have any promotion riding upon this thing," Thackray said. Today a scientist has to attend meetings to be current in his or her field, a participant remarked.

J.J. Berzelius started the annual review, *Jahresberichte* in 1822, to try to give his readers an annual conspectus of what appeared in the literature that year. By that time, most European nations had an academy under royal patronage. Berzelius' work was translated into German by Gmelin, and later by Friedrich Woehler.

Another attempt at creating a premiere journal was the *Comptes Rendus* of the French academy of sciences, which was published weekly in 1835. It was started by members of the academy, to some extent, to answer competition from the popular press, which reported, what was going on within the Academy. It also accepted contributions from nonmembers. "It required an amazing level of organization to get this thing out every week—couriers running all over Paris with proofs," Thackray said. Fifty seamstresses were called in at the last moment to sew the copies together.

EMERGENCE OF PROFESSIONAL SOCIETIES AND THEIR JOURNALS

Science became a profession in the nineteenth century, and professional societies emerged: the British Association for the Advancement of Science in 1831; the American Association for the Advancement of Science, founded in 1848 in Philadelphia by people who had been to the British Association. It was in order to name the people who went to the British Association for the Advancement of Science that the word "scientist" was coined by William Whewell, who had previously come up with the terms "cation," and "anion." This was the moment of professionalization.

Germany was becoming the leading power and locus of professionalization in chemistry. Justus Liebig invented the Ph.D. machine (i.e., the publication machine), Thackray said. Liebig's *Annalen* actually began as the *Annalen der Pharmacie*, because it was with pharmacy students that Liebig began his teaching. In 1832, the name was changed to the *Annalen der Chemie und Pharmacie*. This journal still exists today as the *European Journal of Organic Chemistry*.

This was also the era in which specialist professional societies in chemistry appeared—the Chemical Society of London, the Societé Chimique de Paris, and the Deutsche Chemische Gesellschaft. America was late on the scene. In 1876, the American Chemical Society was a local New York society. The ACS of today really came into existence in the 1890s, 50 years behind the Chemical Society of London. The emergence of professional learned societies led to the development of scholarly journals, funded by societies.

Such efforts were not necessarily easy, because in every case the "turf" was already occupied by other journals. It is instructive to look at development of the *Journal of the American Chemical Society* (*JACS*) as outlined in Figure 2.1. In the beginning, *JACS* was not of much importance. The path to the real *JACS* came by way of the *American Journal of Science*, the *American Chemist,* the *Journal of Physical Chemistry*, the *Journal of Analytical and Applied Chemistry,* and the *American Chemical Journal.*

One scientific society could no longer satisfactorily cover the whole field of research. Thackray provided a quote from J.W. Richards in 1902, "Differentiation and specialization are the watchword, now, of all progress—industrial, scientific, philosophical . . . . The day is past when one scientific society can cover satisfactorily the whole field of scientific research . . . . [T]he analogue of the specialist in science is the society which specializes." That is true of the ACS said Thackray. Around 1902, ACS suddenly discovered

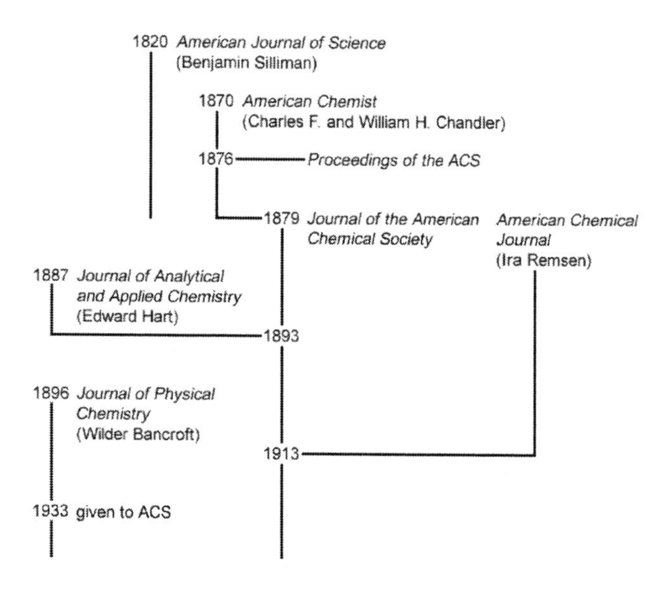

FIGURE 2.1: The path to the real *Journal of the American Chemical Society*. SOURCE: Arnold Thackray and Mary Ellen Bowden, Chemical Heritage Foundation.

that no sooner had it gotten established as a national society than fission occurred. During the decade between 1900 and 1910, numerous other societies were formed, most noticeably the American Institute of Chemical Engineers (AIChE). The ACS began publishing *Industrial & Engineering Chemistry* in 1909, which was a direct response to the formation of AIChE. The stance of the ACS was, "You think you will have a separate organization, do you? No, you won't. We will plant a flag right there," Thackray said.

The American Chemical Society as a broad-based professional society is an extremely interesting phenomenon in the world, Thackray said, because after a wake-up call in that era, it is the society that has most effectively adapted to changing circumstances and kept members inside the "big tent." This, of course, has given ACS some significant clout in the publishing domain, Thackray said.

Another major response to specialization and proliferation is abstracting journals. The first attempt to do this was actually made in 1830, but the twentieth century versions of these journals are more familiar. The original financial plan for abstracting journals was that members' dues would pay for the publications. ACS members, until 1933, received *JACS*, the news edition of *Industrial & Engineering Chemistry*, and *Chemical Abstracts* for their subscription. That model worked for the first quarter-century.

The continuing specialization, proliferation, and growth—chemical databases, Gmelin, Beilstein—was supported by a new frame of thinking in the second half of the nineteenth century, and was driven by technological change. The first essential technology in relation to the scientific enterprise was printing. It was 150 years after the invention of printing (mid-seventeenth century) that the scientific society and the scientific journal were established. In the nineteenth century, the by-products of the Industrial Revolution were revolutions in printing and papermaking. The key change that related to mass literacy was applying the steam engine to the printing press; suddenly it was possible to print many more copies of anything at a reduced cost of production per copy. This revolution fed into the increase in literacy, the rise of the teaching profession, the science teacher, and the appearance of advertising as a financial mechanism that subsidizes journals.

Indicative of this new frame of thinking was the appearance of *Nature* and *Science*, both independent ventures established by entrepreneurial scientists. Norman Lockyer set up *Nature*, and James McKeen Cattell established *Science*. *Nature* remains an independent journal to this day, whereas *Science* eventually formed an alliance with the American Association for the Advancement of Science, but this did not occur until 1944.

This period of time also marked the appearance of explicitly chemical publishers in Britain and in Europe. In the United States, this phenomenon was weaker, in part because the country was still a marginal player, and there was no market for such a tight focus in the United States in the nineteenth century.

THE IDEA OF A "WORLD BRAIN"

By 1900, people were really beginning to feel overwhelmed and besieged by information, and they were looking for ways to put it together. Paul Otlet, Wilhelm Ostwald, and H. G. Wells each had an interesting vision. In 1895, Paul Otlet set up the International Institute of Bibliography in Brussels, subsidized by the Belgian government. His idea was to bring everything together in one place: information from all major libraries and from all publications. He had a large building called the Mundaneum, in which he developed an elaborate version of the Dewey decimal system. He and his co-workers created an enormous catalogue of three-by-five cards carrying information gleaned from the printed catalogues of all the world's great libraries. This enterprise actually survived into the 1970s.

The founder of physical chemistry, Wilhelm Ostwald, took the money from his 1909 Nobel Prize to set up something he called the Bridge. This was supposed to be an office that would be able to answer any question about the literature, the information, and link all organizations working for culture and civilization in the world.

Later, H. G. Wells brought up the idea of a "World Brain." In 1938, he wrote a book of this title, based on the idea of creating a giant encyclopedia of all knowledge. He tried to get Doubleday to take on the project, but it was not interested. Thackray believes that today's Internet is the ultimate expression of Wells' World Brain idea which also fulfills the early missions of Otlet and Ostwald.

JOURNALS SINCE WORLD WAR I

After World War I, the United States emerged onto the chemistry scene. *Chemical Week* began publishing in 1914 in part to promote chemical manufacturing because it was not possible to get fine chemicals from Germany. The social and economic state in wartime and postwar Germany provided an interesting subject for comment by American authors. Thackray quoted William A. Noyes, a great mover and shaker, who said in 1923: "Despite the dreadful financial situation in which Germany finds herself at the present time the *Berichte* [*Berichte der Deutschen Chemischen Gesellschaft*] has published during the past year about 4000 pages of original papers, and this in addition to a large volume of publication in the *Annelen [Justus Liebigs Annalen der Chemie]*, *J. pr. chem.* [*Journal für praktische Chemie*]; *Z. phys. Chem.* [*Zeitschrift für physikalische Chemie*] and other journals. Are we willing to admit that here in America, now the richest country in the World, we can not do as much for our scientific publication as is done by Germany?"[2]

One of the publications that came out of this, at the urging of the National Academy of Sciences, was *Chemical Reviews*.[3] *Annual Reviews* date from this era as well, beginning with the *Annual Review of Biochemistry*.

The era of "Cold War and Hot Science" followed World War II. Thackray discussed the visionary dream of Vannevar Bush in 1945: "Scientific publication has been extended far beyond our present ability to make real use of the record. The means we use for threading through the consequent maze to the momentarily important item is the same as was used in the days of square-rigged ships." In this era, two seminal conferences that began to look at machine searching methods took place; the Royal Society Scientific Information Conference in 1948 and the International Conference on Scientific Information here in the United States in 1958. The computer was there, and the question of, "How are we going to use this thing?" arose. Thackray noted that in the early 1960s the new National Science Foundation (NSF) began made a considerable investment in the development of Chemical Abstracts Service (CAS),[4] noted Thackray (Figure 2.2).

The scene was changing. There was the emergence of the independent entrepreneur—people such as Eugene Garfield founder of ISI (Institute for Scientific Information), who was employed on the *Index Medicus* machine-reading project at the Welch Library. There was the proliferation of the paper record. Europeans were coming in, for example Eric Proskauer, the co-founder of Interscience. Thackray cited a remark of Proskauer's from 1986: "Ostwald would have said that what you need is a textbook to teach the science. You need a journal to publish new developments, [and] you need an encyclopedia as a crowning collection of all the facts."

"So that brings us to where we are today, to the world of complex molecules, complex science, the creation of the Internet, [and] the creation of the wholly on-line journal," Thackray said.

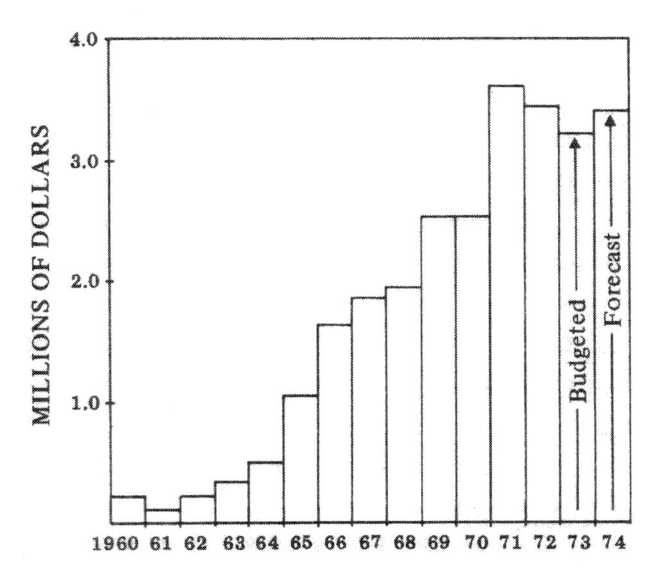

FIGURE 2.2 NSF support of CAS research and development. SOURCE: Proceedings of the Symposium on Chemical Abstracts in Transition, Chicago, IL, August 28, 1973.

[2]Correspondence between William Noyes and Harrison E. Howe, January 5, 1923, Central Policy File, Divisions of the National Research Council, Chemistry and Chemical Technology, Projects, Publication of Chemical Reviews Proposed, 1923, National Academy of Sciences-National Research Council Archives.

[3]M.E. Bowden, "The Early History of Chemical Reviews: 'Established to Fill a Definite Want'" *Chemical Reviews*, 100(1): 13-22 (2000).

[4]The American Chemical Society's CAS indexes and summarizes chemistry-related articles from scientific journals, patents, conference proceedings, and other documents pertinent to chemistry, life sciences, and many other fields.

3

Challenges of Web Publication

The challenges facing chemists and chemical engineers in publishing their work on the Internet were discussed. Preprint servers, authors' rights, distribution, refereeing, search engines, and the amount of material on the web were all topics of this day and a half workshop. Some of the challenges that were mentioned were an increased burden on the reader to find relevant information, the need for special software for some of the enhanced features, adjustments in the publishing processes, the need for a system to tally web hits, a new technology for creating and storing structures, increased investments to meet expectations, and the demands for more rapid and enhanced publishing. A number of existing electronic platforms were discussed.

FINDING INFORMATION

Searching and search engines are ways to find scientific information, but they do not guarantee results. "How do we find relevant information?" Robert Bovenschulte asked. The commercial exploitation of science is a very important factor in chemistry and determines the way ACS approaches the functions it provides. A chemist could, for example, be seeking the latest research on a topic or be exploring developments in an emerging area. Patent attorneys, on the other hand, might need the complete historical coverage of specific reactions and processes.

Bovenschulte discussed a number of web-based search engines and databases that provide content and the tools to access, analyze, and manage research information. Scientists often use Google and Yahoo for first-approximation searches, Bovenschulte said, but these engines deliver too many hits, and not enough specific information. In contrast, Web of Knowledge from Thompson ISI is a much more complete and broad-coverage database with very effective search tools, Bovenschulte added. Chemical Abstracts Service has very broad coverage in chemistry and an interface with many aspects of what has not traditionally been regarded as chemistry. The CAS databases are nearly complete, but they are not perfect according to Bovenschulte. Although scientists have many research tools, using one may not always be easy. "To use it often requires some expert knowledge—or, let's say, to use it expertly requires a great deal of knowledge," Bovenschulte said. STN and SciFinder are other research tools that are available to the scientist. Non-expert searchers more easily use SciFinder, Bovenschulte said. Another ACS product such as STN, by contrast, is much more difficult to use, and one needs some technical training, but it has access to more databases and is favored by those who are doing patent searches, he said.

As search functions become more sophisticated, there is the potential to facilitate the quest for digital information, according to Bovenschulte. Interesting functions such as clustering and taxonomies are being developed and expanded. Search clustering locates articles and can lead to new discoveries, he said. Article linking allows easy access to the full text of cited references. Citation maps provide a visualization of those articles that an individual may be citing, and topic maps provide a way to browse through hierarchies of subjects, Bovenschulte explained.

Advanced approaches to searching are on the horizon—to give some examples: improved filtering of search results based on user profile or user history, automated analysis of document collection, and visual and graphical presentation of search results. Many of these approaches are still very experimental and not widely used because they may require some computer or network reconfiguration. This may act as a barrier because the reader has to decide to invest time and energy, and perhaps some money, Bovenschulte said.

Nonetheless, most of the participants were optimistic about the future of electronic search engines. Stephen Berry said he was very optimistic about the capabilities of enhanced electronic search tools, provided students are trained to use them properly. Wider use of improved search engines can make the publication of scientific papers much more scholarly, because it would be possible to go back and "find who really had the first idea 40 years ago or 60 years ago, not just who published two years ago." Berry pointed to possible future problems due to a changing technical vocabulary, but these are solvable challenges, he said. Different access rules for different journals, variable distribution rules for different kinds of files and the pricing of simultaneous access are other problems the scientific community will face, Berry said.

One of the valuable features in the electronic search process is citation linking. Citation linking is highly desirable for finding older literature, according to Andrea Twiss-Brooks. For engineers, linking should probably not only connect journal articles, but also interweave other literature, such as technical reports and "gray literature." This information is not easy to find either in print or in the electronic world, making the task all the more challenging. Standards for consistent and reliable linking are a major concern. "There is nothing worse than saying that it is out there, here is the link, they click on it, and it doesn't get them to the article," she said. Twiss-Brooks added that there is a need for additional tools, such as a graphical table of contents, to help sort out important information. Table-of-contents services for large journals are not particularly useful, since it does not save time to wade through multiple screens of such tables of contents to find the information being sought.

Structure searching is one of the challenges of chemical e-publishing. "A special problem in chemistry is extracting the science from an article," Bovenschulte said. Although it is possible to incorporate structure searching of articles, it is not easy. Extracting structure information from the submission in digital form is possible. It then can be stored in a standard format and rendered by a viewer. Problems can arise, however, because chemist authors often submit a non-standard view highlighting important features, not realizing that the chemical integrity has unwittingly been sacrificed. Yet if only the chemical information is stored, there is no guarantee that the reader will see exactly what the author intended, and furthermore the reader's viewing software may render information different from the author's original idea. It may therefore be advisable to store a view of the data, as well as the data themselves, Bovenschulte said.

TALLYING WEB HITS

Many participants commented on the need for a tally of web hits or downloads of articles. The two methods that were mentioned were hits per article and COUNTER compliancy. Gordon Hammes said hits per article, which is easily tabulated by all web systems, is the method that will work, but there may be alternative technologies. Patrick Jackson introduced a new initiative called the COUNTER compliancy model that measures the way downloads are tallied and sets certain trade standards. It is essential that downloads are being measured in the same way, he said. Jackson cautioned against tabulating hits alone. COUNTER compliancy levels the playing field, Bovenschulte said. He also encouraged all librarians to begin cost-per-use studies at each site, to determine whether a hit or a download occurred.

DATA

The evolution of electronic journals might spawn new ways of drawing and storing data. At the moment, structures are designed to fit within the standard print environment, according to Carol Carr, managing editor of *Organic Letters*. If there were a way of drawing structures so that "a dendrimer can be a dendrimer, and not have to fit into a one-column space, that would help," Carr said. Carr also said that authors often add color and boxes to their structures, much of which is lost in structure searching. Hammes noted that standard electronic tools in the hands of the authors could simplify the layout process for scientists themselves, giving them control over it, and further contribute to reducing editing and publishing costs.

WEB SUBMISSION

As the web is making journals more accessible to readers, Christopher Reed pointed out that it is also making editors' lives more difficult. On-line submission and other features have made handing in a manuscript a click of a mouse away. Consequently, editors are receiving more manuscripts each year, and are struggling to find appropriate reviewers, Bovenschulte added. Not only are more manuscripts handed in, they often contain more data—crystallographic data or protein database material—so reviewers have to review not only the manuscript, but also the digital interactive images and digital interactive data, he said. This means editors will have to process more manuscripts with more detailed reviews. "One has to worry—and I think it is a very serious worry—that this whole system could collapse of its own weight," Bovenschulte said.

A solution to the growing number of e-submissions and extra data could be automation. The burden could be reduced through automated methods, Bovenschulte said. These would check whether the structure and data files are valid, whether they correspond. There might also be a need to collaborate with other organizations that have greater subject expertise—for example, Cambridge Crystallographic or the National Institute of Standards and Technology (NIST), Bovenschulte said. These connections could be financed through shrinking layout costs. If the print version of a text was to be abandoned and, with it, the need to control page layout carefully, some of the costs associated with publishing would disappear, Bovenschulte said. He said he hopes that there will be continued movement toward a truly seamless, highly integrated electronic publishing process, from the author's submission all the way to the output, whether in print or on the web.

ELECTRONIC PUBLISHING

Giving up print, however, comes with its own set of problems. Martin Blume discussed how some subscribers are concerned that an electronic-only system without an optional print subscription will have nothing to show for it. Being limited to electronic access can make it difficult to keep track of the issues, whereas "If you take print, you always have print," he said. This statement relates to the discussion on access to back issues—journal archives. The American Physical Society (APS) will still arrange for print distribution, but not print, Blume pointed out. Readers will be able to self-print, by using a version of DocuTech, where a file is downloaded, printed, and stored. He added that APS offers a CD collection at the end of the year. Conversion to electronic format for back-files is important because chemists draw substantially on the back-file literature.

Perpetual access will be a question for libraries, according to Twiss-Brooks—especially since the use of e-journals is rapidly increasing. According to a study[1] Twiss-Brooks referred to, "It is not so much a migration from print to electronic as it is a stampede." She also referred to a 1999 study,[2] where it was estimated that one-quarter to one-third of readings came from electronic sources. Not only do current publications have to switch to electronic methods, but there also seems to be a growing need for electronic data repositories. Reed suggested that ACS establish repositories for important data that "need to be added to the storehouse of knowledge, but are not conceptually novel enough or important enough to make a whole piece of paper out of and go through all that process of using the reviewers and everything." This would be an electronic-only repository and therefore cheaper. It would also take away the "bread and butter" of low-quality, high-cost journals, which Reed said he thought, are ruining and exploiting university budgets. This process could also use professional referees, perhaps employees of the professional societies, not practicing scientists. There is no need he said, other than to establish that the work is well done—a peer-review decision on the importance and significance of the work. Several participants agreed with this vision.

However, there was doubt that electronic-only publishing could absolutely guarantee access in the future. Steven Heller, National Institute of Standards and Technology, questioned whether *Chemical Abstracts* had a guarantee of continuing in the future, considering recent advances in technology. "There were a number of abstracting and indexing operations . . . in Germany and England that have disappeared," he said.

Some participants compared the lack of electronic archives in the chemical community to the wealth of preprint archives in the physics world. Berry noted that unrefereed archives started in physics when high-energy physicists circulated preprints, unrefereed for discussion. Berry mentioned how Paul Ginsparg, professor of physics at Cornell University, set up his first preprint archive because it seemed cheaper to do things electronically than to continue to have one full-time secretary in each institution and a very large budget for photo-copying. "The circulation of the preprints went to the 400 and if you weren't a member of the 400, you didn't get one," Berry said. Ginsparg was for democratic circulation, Berry continued. Berry said he thought biology was very conservative with respect to circulating unrefereed articles; whereas physics and astronomy are more open, with chemistry occupying a middle ground—a statement with which a number of participants agreed. Biologists are very concerned that unrefereed articles would be dangerous if the public accessed and used them to try to cure their own diseases, Berry said.

Chemical preprint servers have existed, and some participants pointed out intellectual property and other problems that may arise with the use of preprint servers. Philip Barnett said that there was a small preprint on ChemWeb sponsored by Elsevier, with fewer than a thousand papers, which subsequently died. He thought one reason for this was that some publishers like the ACS do not accept papers for publication that have been in a preprint server. Ned Heindel said he had a citation from an abstract in a regional meeting cited against him as prior art for a patent application. He said this could mean that a preprint is citable as prior art.

There may be additional reasons that other disciplines have been quicker to give open access to their publications. Jeremy Berg pointed to the public as the big driver for NIH policy. He said that many people get health information initially over the Internet and then end up at published articles on research that is often paid for by NIH. When they cannot get access to it, they complain to their representative in Congress, Berg said.

EXISTING ELECTRONIC PLATFORMS

Several participants discussed existing electronic platforms. Jackson talked about the Elsevier platform ScienceDirect. He said it is an easy, stable, intuitive inter-

[1]T. E. Chrzastowski, "Making the Transition from Print to Electronic Serial Collections: A New Model for Academic Chemistry Libraries:" *Journal of the American Society for Information Science and Technology*, 54(12):1141-1148 (2003).

[2]C. M. Brown, "Information Seeking Behavior of Scientists in the Electronic Information Age: Astronomers, Chemists, Mathematicians, and Physicists." *Journal of the American Society for Information Science*, 50(10):929-943 (1999).

face to increase the speed and efficiency of searching. It offers about 24 percent of the world's scientific, technical, and medical (STM) literature on a single platform, and around 30 percent of the world's chemistry literature. Also, Elsevier has pioneered a concept called the "author gateway" that lets authors keep track of their papers from submission to publication. "This actually adds up to what we could call an end-to-end electronic workflow," Jackson said. Not only are authors great submitters in electronic form, but also great users of electronic material, he said.

Bovenschulte cited ACS "ASAP articles"—as soon as publishable—which are generally mounted within 24 to 72 hours after the author has submitted final corrections. However, these efforts require investments on the part of publishers in the whole information technology (IT) infrastructure. The rising number of submissions from around the world has also driven up publication costs, he said.

Michael Keller, Stanford University Press, talked about HighWire Press. HighWire offers HTML and PDF formats, multiple resolutions of images and figures, and easy downloads to citation managers. Many publishers who work with HighWire put up manuscript PDFs on acceptance, and as the articles are edited, they go in to the mainstream. He said faculty members are beginning to use images on HighWire instead of other images; teachers add the URLs for the objects to their course syllabi. HighWire has begun to link articles. For those that are not on-line, a link to a document delivery service is provided. Keller showed some of HighWire's features, like the topic map. The topic map is a graphical navigation device that allows users to move around the 1.8 million full-text articles supported by HighWire and about 15 million articles abstracted in MEDLINE. Keller demonstrated a search: One could choose the search term "genetics" and then sequentially "molecular genetics," "gene regulation," "transcription," "gene expression," and finally "gene networks." A click will pull up a list of relevant articles from MEDLINE and HighWire. A tool called MatchMaker can show the principal ideas, the principal taxonomic terms in the article that created the article's signature. These are the notions that are most important in an article. By clicking on a term—for example, "physiology homeostasis"—the value of that term in the signature is changed, so that a new search can be done for articles with the reweighted term. Keller said there were various limiters—by time, by date, and so forth. Articles in HighWire can be indexed by more than 54,000 terms, which makes it possible to search by idea, rather than just by keyword. He also discussed citation maps and alerts.

4

Cost

Both the cost of producing journals and the cost of acquiring them were topics of discussion. Participants discussed how journal costs could be kept low and how the high prices of journals are affecting colleges, universities, and industry. Several publishers talked about the challenges of pricing their journals and different models of financing journal publication.

THE PRICE TAG FOR BUYING JOURNALS

The cost of journals has risen sharply in the past years, and several participants had ideas of how to bring these costs down. "Solving the problems of expensive journals, and therefore inaccessible to some, will require a very concerted, courageous, and maybe even revolutionary effort by five groups of people," Christopher Reed said. He listed professional societies, librarians, presidents and provosts, scientists, and funding agencies.

Professional societies should better serve the discipline by providing open access as soon as possible, without bankrupting the organization, Reed said. The societies' computers have become the libraries of the world; therefore the societies need the ethics of librarians, not of publishing houses, according to Reed.

Reed believes that libraries are wasting an enormous amount of money binding all copies of a journal. He believes that libraries are becoming white elephants and librarians should teach people how to use the electronic library. He called on presidents and provosts to cut library budgets. "Why are profit-making publishers attracted to this area? Because there is money—if you decrease the amount of money available, we will have less of this," Reed explained.

Reed also said provosts and presidents have to install disincentives for promotion, to discourage faculty from sitting on the editorial advisory board of what he called a "junk journal." Reed asked his fellow scientists to take a pledge not to submit to, referee for, subscribe to, or accept editorial appointments on expensive, non-open-access journals.

Gordon Hammes also asked participants not to support journals that have high costs per page, with the exception of thin journals in important fields that might have a higher cost basis. A number of participants echoed these thoughts. Steven Heller also suggested cutting sales and marketing staff to save money. However, funding agencies could also contribute to lowering journal prices, Reed pointed out, and he listed some ideas to cut costs. Funding agencies should take the publication business more seriously and bear some responsibility for it. There might be a need to explicitly deemphasize the number of publications at grant renewal time, fund for longer periods of time, discourage people from publishing little bits all the time—the least-publishable units—and shorten proposals so that reviewers actually have to read the papers.

The actual cost per user has declined for some journals. The cost per article for the average user has decreased, as a result of increasing expansion and increasing usage, from $12.00 an article in 2001, to $2.00 in 2004, Patrick Jackson said. "We hope to go to $1.00 next year," Jackson continued. He said future chemists would demand that publishers continue to invest heavily in both innovation and quality. Elsevier has data that are currently stored in petabytes, which means the necessary infrastructure is extensive.

Subscription costs per page, however, vary from journal to journal. Hammes compared estimated costs per page for a number of journals and called for uniform page costs to reduce overall library spending. According to Hammes, the *Journal of Biological Chemistry* costs 4 cents per page; *Proceedings of the National Academy of Sciences (PNAS),* is 8 cents per page, and *Biochemistry* is 19 cents per page. *Science* does pretty well at 16 cents per page, Hammes said, but most journals do not carry the same amount of advertising that *Science* does. *Nature* is a very expensive journal at 86 cents per page, according to Hammes. He said that switching to web-only journals would cut costs by about another 20 percent. Combining both of his suggestions—cutting some of the high-cost journals and going online—would leave libraries in pretty good shape.

EFFECTS OF PRICE INCREASES ON COLLEGES AND UNIVERSITIES

According to some participants, the increase in journal prices is forcing many schools to cut subscriptions, resulting in a lower quality of education. Between 1990 and 2000, the budget for science journals in the nation's predominantly undergraduate colleges and universities rose by 120 percent, but this number is deceptive because the increase in journal prices over that period of time was even higher. This means that journals were being eliminated, Michael Doyle said. Martin Apple cited data from the Association of Research Libraries,[1] which showed that journal prices have increased at a rate six times the Consumer Price Index during 1986-2000. Because library budgets are shrinking, and journal and monograph costs are soaring, increasing cancellations must follow, Apple continued. This trend will likely continue into the future as library budgets continue to shrink, he said. State universities especially face severe problems. An increase in future state and federal budgets for libraries, research, or other educational expenses should not be expected. "In fact, before the House and Senate right now is the first cut in the National Science Foundation, practically since it was founded," Apple said.

Some publishers do offer discounts to schools. Martin Blume said that the American Physical Society offers online-only for 15 percent less for larger institutions and 20 percent less for smaller institutions. However, for some institutions, this may not be enough. Grace Baysinger said that many of the small schools could not afford Chemical Abstracts Service (CAS), even in consortium agreements, which might very well have an impact on tomorrow's researchers. Baysinger reminded participants that 50 percent of Ph.D. students come from small schools, most of which have very limited access to CAS. The ability to find information is critical, and the ability of the discipline to support access to the major database is important, Baysinger pointed out.

Dennis Chamot drew a parallel to health care. Reed agreed, and said that a recent book suggests that 20 years ago, health care costs for administration were about 10 percent. After privatization, they are now about 17 percent. "So the profit motive adds to the cost of health care and doesn't make it any more accessible," according to Reed. He said that this is why he thought profit makers do not deserve to be in the business.

[1]For more explanation of the data and more recent statistical analysis, see the Association of Research Libraries web site at www.arl.org.

THE ROLE OF PROFESSIONAL SOCIETIES

Nicholas Cozzarelli said he did not think that librarians really want to support all of the activities of the American Chemical Society or the American Society of Biochemistry and Molecular Biology, but rather they want to buy that society's journal. He thinks the system has brought about some excesses and listed the salaries of ACS officers as an illustration. Charles Casey spoke for himself (as the 2004 ACS president) and said that he was lobbying for more openness and internally trying to get some moderation in ACS salaries.

Apple pointed out that most societies in the Council of Scientific Society Presidents (CSSP) barely break even on their journals. The average loss per year is about $200,000, and they cannot shoulder additional burdens, he said. The science association publishers' competitive advantage is higher quality. In cost per citation to the library, the difference may be ten- to fifteen-fold in purchasing from association journals versus commercially published journals, Apple said.

Michael Keller said that HighWire Press can help publishers, especially not-for-profit publishers, compete more effectively. HighWire serves 125 publishers, mostly from scientific societies and a few for-profit publishers. He said that HighWire Press was established to do two things: (1) to use network technology to enhance scholarly communication, and (2) to make not-for-profit society publishers more competitive.

Publishing is a very large commercial market; in 2002 it was about $7 billion according to data that Apple showed from the Electronic Publishing Services (EPS) Limited. "Because it is a big market, that makes a lot of difference to a lot of people about where it should go and what it should do," Apple explained. Elsevier accounts for about a quarter of the entire market, he said. The top four providers account for about half of the market, and there are only a couple of major professional societies among the top 15—the American Chemical Society and the Institute of Electrical and Electronics Engineers (IEEE). He said it should be recognized that the societies are in the business of helping develop the next generation of scientists, as opposed to making profits for shareholders.

MODELS FOR FINANCING SCIENTIFIC JOURNALS

There are several different payment models for financing journal publication—library pays or subscription model, author pays, pay per view, and a combination of any of the above, possibly with page or color charges. Participants voiced their opinions on the strengths and weaknesses of each and discussed what a new model could mean for their institutions.

Subscription-Based Model

The most prevalent payment model is "library pays" with no extra page or color charges to the authors, Reed said. He did not agree with the common notion that the indirect funders—the government and research foundations—really pay for these costs. On the contrary, indirect costs for libraries are capped, he said. The institutions pay out of funds that could be going to research and education, according to Reed. However, subscriptions come with an extra cost for publishers, Cozzarelli said. *PNAS* loses about a third of all subscription revenue to subscription agents and other intermediaries, a fact which makes subscriptions financially inefficient. The subscriber model is market driven, Peter Gregory said. Gregory listed some of the consequences of subscription models such as centralized buying, a stress on library funding, and the expectation that industry pays it way. One of the strengths of the subscriber model in comparison to the "author-pays" model is that publication cannot be bought (i.e., as if it were an advertisement), Gregory said. A combination of library pays with additional page and color charges to the author constitutes the second payment possibility.

Pay per View

Pay per view is yet another model of financing journals. This model has its own pitfalls. Publishers might emphasize content that is likely to be highly viewed, and dull research might not be published, Gregory said. "And the consequence of that, again, would be that scientists and chemists especially would be repeating the dull but correct stuff again and again because no one bothers to publish it," he said. The pay-per-view model is useful only as an additional service, Gregory concluded. He said it is far more important that the publication is based on the ability to do science rather than the ability to pay to publish the results.

Author Pays

Another model for financing journal publication is an up-front payment by authors, but problems arise with this model: wealthy research groups plead poverty and instead use their funds for other expenses. These charging models have failed in chemistry and physics in the past, Gregory said. "The authors don't want to pay," he said. There are also ethical problems with an "author-pays" model. "If there is money involved, either with referees or with authors who can pay to be published, it is more likely from our editorial experiences that there is corruption waiting just around the corner," Gregory said. Shifting the cost from libraries to authors might make libraries obsolete, and industry would be the net beneficiary of this model, he said. This might mean a loss of income for some publishers. The Royal Society of Chemistry, for example, derives around 45 percent of its income from industry, which would have to be found from other sources if author pays were the only model. Authors themselves might also have to scramble for money if the payment model were to be changed, because Gregory doubted that enough academic funding is in place to support the system. Administrative and billing problems might also arise; publishers would be dealing with every author, not just with a few thousand customers. Problems could arise for both publishers and universities. Publishers do not have the means to collect outstanding author fee debt, and universities would have to deal with publishers from all over the world in all different currencies.

There are some problems with an author-pays model in the context of open-access publishing, Gregory added. He fully agrees with the idea of complete access to information, but cautioned that unclear financing could irreparably damage the great heritage of the American Chemical Society, the Royal Society of Chemistry, and chemistry itself over hundreds of years. Stevan Harnad objected to discussing "open access" in the context of publication financing. He said that no government is mandating an author-pays model, but added that the United States, the United Kingdom, Canada, Australia, India, Brazil, Norway, Denmark, and a few other countries are recommending open-access models—where articles are freely available to anyone on-line immediately upon publication—not author-pays models.

Gregory replied that the U.K. research council's draft proposal includes a requirement for 2004 in which author-pays models must be are strongly considered by all research councils. The question was raised whether the British response was guided by the fact that the publication industry there is more dependent on overseas subscriptions perhaps than in other countries. According to Gregory, a good deal of the science and technical publishing industry is based in or has major units in the United Kingdom. A high percentage of the Royal Society of Chemistry's revenues come from overseas—about 85 percent—representing a significant injection of cash for U.K. science. He said that this revenue benefits the UK trade balance, UK employment, chemical sciences worldwide (because the RSC acts worldwide), and is overall good for the UK contribution to developing and supporting the chemical sciences. "But the main winner is the chemical sciences wherever they are, as the money is spent on science rather than on commercial publisher shareholders," Gregory added.

A number of participants related their experiences with authors' not paying page charges. Cozzarelli said that this was not a problem at *PNAS*. On the other hand, ACS did have a problem before abandoning the author-pays model. *Physical Review Letters* has voluntary page charges, but other APS journals do not. Blume said that authors choose where to publish, and they will choose a journal without charges if they can. However, he pointed out that asking an author to pay so that there will be open access might change this response. That is different from asking for page charges where there are controls, which was the situation in the past, Blume said.

Cozzarelli added that money had no influence on the papers he published. Gregory recounted having had money offered to him from a corporation that wanted to avoid patent issues to publish a paper. Hammes thought it appropriate that institutions pay a substantial part of the cost, because publication furthers the institution, whose business it is to get the research out there. Stephen Berry said that the reason a federal agency or the not-for-profits support research is because it will generate a public good. A public good is an item that is not diminished in value by use, he said. Scientific public goods are special because they increase in value with use. Agencies that support research should be prepared to pay for publication, he said. Berry called for an economic analysis of all the plausible modes of supporting journals to determine which have the lowest transaction costs, and where the largest fraction of the money is spent directly in supporting the publication. He said that page charges may look good, but they have large transaction costs because of the overhead steps from start to finish. Gregory said that researchers might not have a research grant. "How are you going to pay from your nonexistent research grant a publication fee?" he said.

According to Leah Solla, Cornell University investigated what would happen to its costs in an open-access environment.[2] Calculations were based on how much the library spends on subscription costs and how many articles Cornell authors publish. The study found that Cornell spends $1,100 for every article published by a Cornell first author. If larger commercial publishers are removed from that equation—Elsevier, Wiley, and Springer, which make up about 25 percent of the articles and about three-quarters of expenditures—the cost drops to about $400 per article. Cornell did not research all author charges or all personal subscriptions of all of the researchers on campus, but the study still found the library-pays model to be more favorable for the university. "I don't think that that would nearly add up to $3,000 or $4,000, the kinds of amounts that we have been hearing about today," Solla said. Cornell would not easily adopt the author-pays model, she said. An author-pays model would shift costs to larger universities, but is not clear who would benefit from this shift. Reed cited the *Journal of Financial Economics*, which charges for submissions and then reimburses authors for accepted papers, as a pricing model.

Although a shift in payment models may mean larger contributions for some institutions, they might benefit from other aspects of the change. Undoubtedly some institutions will wind up having a slightly larger contribution, Vivian Siegel said. Yet in looking at the outcome, those institutions then also have access to large degrees of information that they might not otherwise have. Focusing entirely on the library budget is really missing the point of what a change is all about, she added.

An author-pays model might deter researchers working in the chemical industry, said Parry Norling, who spent 35 years at the DuPont Company. He said that the chemical industry is often reluctant to publish. While the patents that scientists develop are paid for by the company, page charges for research articles might have to be paid out of the scientists' own pockets, unless they could argue that there is a real benefit to the company. He summed up his experience at DuPont by saying that page charges are a barrier to publishing itself, not only a factor in the choice of journal.

THE COST OF ARCHIVING

One of the costs that open access would incur is the cost of archiving. Hammes talked about models to pay for keeping an open archive. He said that raising the subscription price of a journal slightly would amortize the cost of archiving in a very short time. He urged societies that do not follow this model to make their archives free now and solve a lot of accessibility questions in the process. Andrea Twiss-Brooks said that such models would aid small institutions that may not be able to afford all of the on-line subscriptions for current periodicals. Patrick Jackson, however, explained that Elsevier could not give open access to its archives, because the company had invested $40 million in them and needed to earn back the investment.

Hammes said that the *Journal of Biological Chemistry* has been giving free access to its files after six months for several years and has not lost money. He added that the number of subscriptions has gone down, but in the same way they have for other journals, as purchasers eliminate duplicate subscriptions. The ACS archive is used heavily, but not nearly as heavily as current subscriptions are, according to Twiss-Brooks. She said that making the archives freely available would not impact current subscriptions. Some participants, however, felt that giving access too quickly might damage publishers. Blume said the 28 percent of APS subscriptions from the smallest institutions are the most vulnerable in the event APS makes everything open access very quickly. With them, there is a potential revenue loss of 30 percent, and possibly more. "That is why we are cautious about this," Blume said. There is a possibility of granting open access for the entire archive now, if the subscribers sponsor it, continued to sponsor it, and agreed to increase their contributions in the future. Blume added that two APS journals are now open access. The first, *Physical Review Special Topics—Accelerators and Beams*, is supported by institutional sponsorship. Charging the author or the author's

---

[2]Available from the Cornell University institutional repository: *http://techreports.library.cornell.edu:8081/Dienst/UI/1.0/Display/cul.lib/2004-3.*

institution about $1,000 an article will support the upcoming *Special Topics—Physics Education Research.*

Jackson said that in general if a chemistry customer buys the back-files from before 1995, statistics indicate that the usage is usually around 15 percent, a number he thinks is significant.

AN INDUSTRY LIBRARIAN'S APPROACH TO ACQUIRING CONTENT

Lou Ann Di Nallo explained Bristol-Myers Squibb's (BMS) approach to acquiring content. Her company licenses content globally so that it is available to its researchers no matter where they are, and trains its employees to use the electronic scientific literature to leverage its investment. Her company works through the purchasing organization Global Strategic Sourcing to obtain access to these journals though, which can be a challenge. "There is a little bit of a learning curve there for them to understand and accept that there are not four different companies out there waiting to sell you access to *Tetrahedron*, there is one and you have to deal with that," she said.

According to Di Nallo, the BMS library's strategy is often more of cost avoidance than cost savings, which has a lot to do with BMS being a very electronic environment. When, some years ago, there were significant budgetary pressures, the company canceled a lot of print, Di Nallo said. After an uproar by some researchers, some titles were reinstated, but after a year, statistics showed that print was not used heavily. The library keeps print from publishers who discount print with the electronic access. "The only reason we have the print is because it was actually at a discount," Di Nallo said. She said the scientists are not really aware of the actual costs involved, just for the print. Added to this is the cost of having someone there to receive the print, to check it in, and to put it on the shelves, she said.

BMS continually reviews requests for new content through a couple of mechanisms. One is a user community of about 75 people representing different areas within the company. Then there is the Content Advisory Group, which is made up of senior-level members of the research institute. Around budget time, the company looks to this group to help it make difficult decisions. Di Nallo added that her company is almost obsessive about usage, statistics, and other metrics. Cost per use is a factor that guides budget decisions and negotiations with publishers. She said that this is not the only measure of value, but it is a pretty good one and something to which finance people can certainly relate. She said that the journals are expensive, but the cost of the journals pales in comparison to what BMS pays for the tools to access journal literature.

Di Nallo said that one of her challenges is to get the most out of her budget as content costs are rising. "There is a real balancing act that goes on there trying to figure out how to squeeze the most 'bang for your buck' out of the overall budget," she said. Di Nallo said she felt that less competition in the publishing industry is leading to higher prices, and she may have to deal with this by limiting access. "We have been getting the word out there to our users about these journals. We have been putting linking solutions in place. We have been doing training education, and now we are really finding ourselves in the very awkward and uncomfortable place of having to think about limiting access to some of the stuff," she said. She added that this has resulted in her library paying twice as much. Di Nallo also added that employee education is becoming a large component of her work, not only on what is available and how to use the tools, but also on the costs. "Typically, what I am finding is they are less inclined to pay for things than librarians are, once they are aware of the actual cost," she said.

COST STRUCTURE OF STM JOURNALS

Editors from the Royal Society of Chemistry, the American Physical Society and the Public Library of Science, shared details and challenges of the cost structure of their publications with participants.

Royal Society of Chemistry

Several factors drive the cost of publishing a journal, Gregory said. Inflation, attrition, a publisher's investment in its electronic platforms such as ScienceDirect, Wiley Interscience, or those of the ACS or the Royal Society of Chemistry, and the loss of subscriptions all contribute to rising prices for the Royal Society of Chemistry's journals. Increased submissions and increased publishing output also drive the pricing considerations of all publishers, Gregory said. The number of the Royal Society of Chemistry's core journal submissions increased by 17 percent from 2002 to 2003. This translates to 17 percent more work for the staff, or more staff, he said. "The question of the rejection rate came up early, high ones, low ones; who does [the review]?" he said. The Royal Society of Chemistry has 70 chemists dedicated to conducting the peer-review process. It rejects about 25 percent of submissions without even putting them out to other peers. The result of increased rejection rates for massively increased submissions is 6.5 percent more output from Royal Society journals in the period 2003-2004, he said. The European Community adds a 17.5 percent VAT (value-added tax) bill to electronic services, he said. As a consequence, the Royal Society of Chemistry supplies its customers with print, because many are actually buying print to avoid paying the VAT.

American Physical Society

Blume talked about the economics of American Physical Society publishing. As with the Royal Society of Chemistry, the number of submitted and published articles is the basic factor that drives cost, coupled with inflation, Blume said. It often costs more to reject an article, in editor's time, than to accept it, he said. Other costs include the office, telephone, postage, composition, production, and distribution. These costs are important, Blume said. Distribution costs are now for both electronic and print with electronic-only leading to significant savings. Active subscribers are about 20 percent electronic-only, not counting consortium agreements, which give electronic free distribution, he said. According to Blume, the net effect of dropping print varies is a cost reduction of 15 to 20 percent, or more in Europe, because subscribers not longer have to pay airfreight.

Blume compared the cost of the APS-all package—which encompasses *Physical Review A, B, C, D, E, Physical Review Letters*, and *Reviews of Modern Physics*—for a particular large institution in 1988 and 2004. In 1998, APS-all cost $12,015. In 2004, the cost increased to $24,570, with smaller institutions paying less than larger institutions due to newly introduced tiered pricing. However, this institution actually paid less to access the ACS-all package in 2004 than 1998, because in 1988 it held three APS-all subscriptions—one additional subscription to *Physical Review Letters*, and one additional subscription to *Reviews of Modern Physics*, whereas in 2004, the institute simply had one APS-all subscription. Factoring in the duplicate subscriptions, the total cost for the institute in 1998 was actually $38,300. In 1998, print was included with 90,000 pages; in 2004, it was print with 110,000 pages. "You would think that this would be enough to tell people to not just look at the prices, but look at what you get for it at the same time and what has happened in the interim. The large institutions have saved heavily," he said. The cost per article for large institutions is 16 cents a page. APS tries to price its journals so that the cost per page is the same across all journals.

As for other publishers, rising submissions drive prices. Last year, APS received 27,000 submissions, which requires a significant editorial staff. APS has an in-house staff of 35 editors plus about 50 editors based around the world—so-called remote editors like Jack Sandweiss at Yale, the editor-in-chief for *Physical Review Letters*. There is a widening gap between the number of published articles and the number of submitted articles, Blume said. He said that the average cost of an article is $2,000. Eliminating print would lower this even more. Blume said that APS is lowering prices next year because of the benefits from electronic work in the office.

Public Library of Science

Siegel talked about the Public Library of Science (PLoS) and its business model. This model involves recovering all of the costs of publication (including peer review, production through the on-line version, and all associated overhead costs) through up-front publication charges. Up-front charges are currently set at $1500 per submission.

According to Siegel, PLoS thinks that up-front charges should cover publication through the on-line version, and it then sells print versions of its journals at the cost of printing and circulation, supplemented by print advertising. When a researcher submits an article, he or she indicates what part of the charge the research can pay. This information is shielded from anyone who can make a decision about what to publish. Currently, slightly less than 5 percent of the total publication revenue is lost to nonpaying or partially paying authors, Siegel noted. There are no additional charges, no color charges, and no individual page charges. There is no arbitrary size limit to any of the papers. PLoS has several grants through the startup period. It also has membership programs, and allows corporate sponsors, albeit very carefully, Siegel said.

5

Access

Do chemists have adequate access to the quantity and quality of journals they need? Participants looked at the situation for developing countries, students, industry, and chemistry in general, and at the unique publishing approach of the *Journal of Atmospheric Chemistry and Physics*.

THE DEVELOPING WORLD

The number of people around the world who have access to content on the web has increased in the last few years. Because the content of almost all scientific publishers, whether society or commercial is available on the Internet, Robert Bovenschulte pointed out. "Everyone is publishing the material on the web a short time after the authors have submitted final corrections," he said.

Licenses have also increased access, Bovenschulte said. The ACS embarked on licensing arrangements around the world—for example, with some 80 institutions in China. The ACS is starting experiments providing access to developing countries that do not have the financial means to pay for the content.

There still may be financial barriers to access, however. Bovenschulte estimated that 30 percent of potential users worldwide of the full text of American Chemical Society journals cannot access them today because their institutions cannot afford the fees. He added that he has seen such rapid progress in expanding availability in the past few years that he believed the number would come down. In looking only at the number of scientists who are very actively working at the higher levels of research in a given field, those who lack full access might be lower. He added that given the arrangements that ACS has throughout the world—all of Brazil has a consortium arrangement—virtually everyone who needs access to ACS journals has it, because of the spread of the consortia arrangements.

Yet even these numbers are too low for some authors. Stevan Harnad said that from the author's point of view, they are losing potential impact by not reaching out to readers who lack access; he thinks that the ACS should adopt a policy that would welcome the authors' making their own articles available on-line to would-be users for free.

According to Bridget Coughlin, *PNAS* access is free to developing countries that are working on building their scientific infrastructure, including China, Mexico, all of Latin America, and Africa. She said that *PNAS* participates in the American Chemical Society's Bookshare program, as well as eIFL (Electronic Information for Libraries), HINARI (Health InterNetwork Access to Research Initiative), and other initiatives to distribute electronic scientific information globally.

AMERICAN UNIVERSITIES

Small universities and undergraduate institutions may have problems accessing some journals, whereas larger universities with bigger budgets appear to have enough access. Christopher Reed said he is at an institution that has an adequate library budget; the journals are not too expensive for him and they are quite accessible. "But I think that is only because I am at a big institution that really can afford to pay," he said. He added he would like to see the literature free anywhere, at any computer terminal, any time, in the world.

There is much at stake for chemistry as a discipline. There is concern about the ability to find information. People should be able to access the information that has been created to maintain the health of chemistry as a discipline, Grace Baysinger said. Young scientists are the seed corn of the profession, Reed said. They see it as a noble, vibrant, and important discipline. "We must make access available to them, instead of getting caught up in all this market stuff that is preventing it," Reed said.

Publishers offer special packages to universities. The American Chemical Society recognizes the financial plight of some needy institutions. ACS offers educational packages, journals at a lower cost. Other institutions such as California State Fullerton or Williams College reduced their number of chemistry journals from 70-80 in 1990 to 60 in 2000, Michael Doyle said. However, there are also institutions like Ohio Wesleyan and other very small institutions that may have only 15 journals which they are struggling to maintain. He said one size does not fit all. There are some very high quality undergraduate institutions that spend much more than others on their library budget, have much greater access to instrumentation and hold funded research grants. There are other institutions that are struggling desperately.

Some workshop participants saw parallels between developing countries and poor universities. Doyle compared the plight of predominantly undergraduate institutions to the different tiers of developing countries. Some are doing very well and have more than a billion dollars for a small number of students, he said. Others are really very poor. "If you don't have access to certain journals, you can't remain very current," he said. If a student or researcher has to go a hundred miles to reach the nearest institution with a major library, that is a limitation. The pricing of journals impacts how this community reacts and will shape the educational well-being of the students.

CHEMISTS AND CHEMICAL ENGINEERS

After looking at developing countries, and educational institutions, the participants discussed whether chemists and chemical engineers had adequate access to literature.

Access (via subscription or otherwise) was generally considered sufficient by many of the workshop participants. Today's chemists have more access to more content and functionality than ever before, Patrick Jackson said. The term "gray literature" is being used to mean those journals that are not on-line. Many believe that publishers who have not made their journals electronic may be marginalizing their content. Print has limited accessibility, and this favors going to completely electronic media, Gordon Hammes said. Usage and access to the literature have never been greater, according to Peter Gregory. Increasing access to scientific information is crucial. The Royal Society is very much dedicated to this and is open to as many ways of doing it as possible, Gregory said.

Yet not all access situations are ideal. Many readers cannot view all the research they want when they want it. Clinical medical research studies are one example, and Martin Apple said he does not think it is justifiable to keep them withheld. Ulrich Pöschl said that the Elsevier slogan "access equals impact equals value" is flawed. He said you must multiply impact by quality to get value.

Access to data may be crucial in the future. Apple said that the scientific community should pay a lot more attention to the availability of data, even more so than to journal access. A responsible scientific society should not charge for its archives. Free archives would cover most accessibility questions. The scientific community "can't live without data and we can't live without access to data, and I think this is something that we really need to pay a lot more attention to," Stephen Berry added.

Access for industry scientists was also deemed sufficient. Chemists and chemical engineers are getting the needed access at Bristol-Myers Squibb, Lou Ann Di Nallo said. Return-on-investment clearly drives resource allocations, and there is a big focus on how money is spent in the industry library. This means that increasing prices might lead companies to further limit access to some journals in the future.

The *Journal of Atmospheric Chemistry and Physics*

Ulrich Pöschl talked about the *Journal of Atmospheric Chemistry and Physics*, an on-line open-access journal that ensures quality through interactive peer review and public discussion with a two-stage publication process. In Pöschl's opinion, open-access publishing can and will substantially improve scientific communication and quality assurance. The *Journal of Atmospheric Chemistry and Physics* was launched three years ago. Pöschl's focus is on the improvement of scientific quality assurance. The shortcomings in the closed peer-review process are twofold. Critical messages are watered down, and the review process presents an opportunity for hidden plagiarism. Although traditional peer review works very well in many instances, very valuable comments made by the referees are lost. This leads to a decrease of scientific discussion in traditional journals.

There are two conflicting needs in publishing: (1) rapid publication and (2) thorough review and discussion. Pöschl explained that a two-stage publication process with full traditional peer review can meet both of these needs. In this process, what Pöschl calls a discussion paper is published rapidly, a kind of upgrade preprint. If the paper merits review, it undergoes full peer review and public discussion, and referees publish their comments alongside the paper, anonymously if they want to. Additional comments from interested colleagues are also published on the Internet, the discussion is closed, and the peer review is completed, at which stage a paper can still be rejected; then a final revised paper results, as in a traditional journal. This can be done in ten days to eight weeks, after which traditional peer review and final publication take three to six months.

Authors, referees, and readers gain from this model, Pöschl said. The discussion paper offers free speech and rapid publication to the authors. Interactive peer review and public discussion allow direct feedback and public recognition of high-quality papers. There is less time for obstruction and plagiarism. Critical comments, controversial arguments, and both scientific flaws and controversial innovations are documented in the discussion paper, Pöschl said. He added that public discussion deters careless and useless papers. The journal is fully covered by ISI and Chemical Abstracts Service from the first paper on, and it publishes 300 papers per year. Submission rates are increasing at about 20 papers per month; the rejection rate is 10 percent. The impact factor is 2.3, which is among the highest-quality atmospheric science journals. The journal is first among atmospheric science journals in the immediacy index, a measure of how many papers are cited in the year they are published.

Pöschl talked about some statistics of the discussion forums. There are about four comments per paper on average, most of which are actually refereed comments and author responses. One in four papers is additionally commented on; for traditional papers, the number is one in a hundred. The possibility of public comment is a motivation to turn in a good paper, and results in a low rejection rate. Public comments have to be signed with the commentator's name. Although some comments are harsh, there has not been a case of personal offense. The comments, from both the public and the referees, are archived and fully citable. The second journal based on this model has been published, *Biogeosciences*. The page charge is 20 euros per page. Single issues are printed on demand and sold for 6 euros per issue.

Pöschl said that publishing on traditional preprint servers and then in traditional journals is less than optimal because the opportunity for discussion is lost. An idea for the future might be to split the papers into different categories of different scientific value. His model could be an intermediate between the traditional journal and full open access. His ultimate vision is open-access publishing, peer review and discussion to obtain better and fewer papers, but he aims for more than preprint self-archiving and impact. "I want to see improved scientific quality, and for this I need open access with interactive means of the Internet, interactive discussion," he said.

There is a relationship to the physics archive, since it is possible to print comments on papers in the physics archive; the comments appear to be regular and are linked to the other papers, Martin Blume said. He added that all versions of the papers are retained in the archive.

6

Archives

Participants discussed copyright issues, databases, repositories, and methods to retain the value of the scientific work.

Archives are a key component in the change from print to electronic publishing. The position of the ACS Publications Division is that it is prepared to stop having printed journals when two changes occur. One is when authors decide print no longer adds value, Robert Bovenschulte said. The other is when there is a reliable technology available to store the archives, so that the science will be preserved, even without print. Currently, print is much more reliable than the current technology. Martin Apple has observed that there is worldwide scholarly journal dissatisfaction for many reasons; including the pervasive reluctance of libraries to cancel print until e-archiving arrangements are secure and durable. Another reason is the need for organizational structures to ensure access to digital archives.

Maintaining archives in the future is one of the main problems. There is a pressing question of how to maintain these functions over time, especially in the eye of technology migration. "How do we take electronic technologies that exist today and, if we embody our content into those, what will happen 5 years from now, 10 years from now, 20 years from now?" Bovenschulte asked. To the extent that publishers like the ACS feel it is their professional obligation to enable such migration, they have to invest a great deal to make sure that they can handle these transitions, he added.

Some commercial publishers have already started to digitize their back issues. Patrick Jackson said that *ScienceDirect* today has about 2 million current articles and about 4 million articles that belong to the back-files. These are increasingly being supplemented with other types of content, including reference works, books, and handbooks. Material that was published 50 years ago, or even 100 years ago, is in many cases held to be just as valid now as it was then. Elsevier began to digitize all of its journals around the year 2000 and will be finished by 2005, at a cost of $40 million. All 1,800 Elsevier titles, including all of the discontinued, split, and name-change journals will be included. Elsevier sells about 29 different back-file packages. The key benefits are getting rid of the physical archive, access to internal links and CrossRef links, and free access to 6 million abstracts. In case disaster strikes there is a contingency. Comparing paper to electronic files in terms of reliability, Jackson reminded listeners of what can happen to libraries in times of war (e.g., Bosnian National Library) and natural disasters (e.g., earthquake, Kobe, Japan).

Not all costs can be cut, however. Gordon Hammes said that data archiving could possibly be done more efficiently, but the data still have to be reviewed. Not all data should be archived, or archives of incorrect or obsolete data would ensue.

Martin Blume said that the American Physical Society (APS) offers a CD collection at the end of the year that can be loaded on the intranet of the institution, so that everyone will have desktop access to it. There has been a call in the United Kingdom for every researcher to self-archive every published article in a peer-review journal in his institutional archives, Stevan Harnad said.

Participants discussed some copyright issues that electronic archives might create. There was a policy forum for science some years ago arguing for authors to retain copyright the way novelists do, Stephen Berry said. "Every scientist gives the publishers the copyright, and one way that has worked very well is for the journal that holds the copyright to give the author a very, very open license," Berry said. The APS uses this model. The only reason left to argue over copyright ownership is the original intent of giving the author or inventor the protection that comes from being the creator, according to Berry. Although the issues of intellectual property rights are quite central, they are solvable in our context, Berry maintained. He called for an analysis of the financial pathways to open access, but cautioned to not expect any one simple solution.

Alternative models to existing copyright laws already exist—Creative Commons being one of those that the participants discussed. Creative Commons is an important alternative to the conventional understanding of copyright, Anna Gold said. This initiative originated in the artistic, creative part of our culture and has moved into the area of scientific creativity. Creative Commons provides well-crafted legal agreements that allow authors to both control their creative property and enable its reuse without having to be contacted themselves, so they can maintain control while providing access.

Berry talked about the pressing problem of databases and the laws protecting the data, especially in the European Union. The European Union "database directive" created a specific new kind of protection for databases that is more protective than even a copyright. As a result, a number of privately owned and distributed databases in Europe have been created, many of which are very expensive. There was one attempt in the United States to protect satellite data in this way. That idea failed because no one in space science could afford the data. There is an ongoing battle in Congress about whether to enact a law comparable to the one in Europe, Berry said. However, most scientific data are the kind of raw data that can be copyrighted. The *Journal of Physical and Chemical Reference Data*, for example, is copyrighted because it contains evaluated data. Berry doubted whether truly raw data still exists from scientific experiments. "When we do an experiment in my lab, we do not simply collect the electrical impulses that detectors find and print those out or put them directly onto some electronic record," Berry said.

DATA REGISTRIES

Berry also discussed how a data registry might be needed in the chemistry community—a registry that would not be a data repository, but merely information about whether and where data exist. He said, there could be a great amount of interest in a global biologicals registry, but this might be difficult to set up because many developing countries are very protective of their native flora inventory data. They fear they will be exploited, so they are very cautious about allowing people to construct databases of such information. The chemistry community has large sectors in which work is done on potential pharmaceuticals of natural products, where having data repositories of substances recovered from organisms but not yet studied would be very valuable, according to Berry. Bristol-Myers Squibb would probably be interested in accessing such a registry, but probably would not contribute to it because a registry is a company's intellectual capital and money, Lou Ann Di Nallo said. According to Ned Heindel, this concept is not at all new. He said the ACS had a section in its *Journal of Medicinal Chemistry* some years ago that listed negative results for "me-too" compounds. Di Nallo added that it is extremely helpful to the pharmaceutical industry to know which compounds have no activity.

THE FUTURE OF ARCHIVES

Gold talked about the future of archives and some archival tools. The journal is not the final stop on the scientific communication road; the archive is, Gold said. The archive is never final, it is a way to preserve scientific knowledge, to preserve the record so that it can be built on and used into the future.

Some of the problems of archiving include reliability, Gold said. Librarians and scholars have dealt subsequently with publishers who have left the scene and dealt with how to recover data, records, and so forth. Reliable archives will benefit our children and our grandchildren, but in the digital realm, reliability into the future is not a foregone conclusion, she said.

The archive is important because it provides context for work—not merely a way of getting at a particular known piece of work. Libraries provide that context by bringing together the patchwork of various publishers and models, and then deal with the frustrations of trying to piece it all together. Libraries work toward a grand vision of a richer and more interoperable context, Gold said.

Some of the solutions in terms of reliability include finding ways to agree on and share responsibility, according to Gold, and cannot be done in a single organization. Open access with cost sharing in some way and Creative Commons as a means of managing access are very promising ways of handling intellectual property issues and dealing with management and governance issues to help us move into the future. Gold named some current approaches, such as *JSTOR*, *DSpace*, and *E-Depot*. *JSTOR* is a multi-institution approach to providing access to a historical journal archive. *E-Depot* is a national-library-plus-publisher initiative to ensure the longevity and reliability of a digital archive into the future. *DSpace* is MIT Libraries' multi-institution federation. It is institutional repository software, but also preservation repository software, intended to be open-source and openly developed across many institutions. Possible content ranges across the spectrum from journals to many other kinds.

Some participants felt that depositing is an important part of archiving. The feeling was expressed that either the process of depositing should be part of a seamless workflow, which might be automated through harvesting, or it has to be stewarded. To leave this to individual faculty or their administrative assistants, some felt, is extremely unreliable.

Although there is a tendency in e-business and the Internet-enabled communications industry to charge very little or zero for content, this does not rule out charging for other value added, Gold said. Open access begins to allow scientists to work with their archive in much more creative ways. She cited some very interesting recent articles[1] about how the archive could actually begin to represent the dynamics of scholarship in more creative ways than it does now.

A new system will mean new archival choices and challenges, Gold said. The perspective of archivists is more and more becoming a core part of what libraries do throughout their organization. Archivists help preserve the context, the dynamics of knowledge, and libraries will begin to play a much greater role in this activity, she noted. The challenges for providing this kind of dynamic archive are immense as well: interoperability, selecting information into the archive, managing an archive in environments where people have little time. She added that developed and agreed-on standards that could support such an interoperable world and the migration of functionality over time are further challenges.

According to Gold, chemistry has much at stake. Chemists are heavy users of journals, their journals are generally agreed to be the most expensive, and costly and complex data are embedded in their literature, she said. Opportunities may be lost: interoperability with related disciplines, interoperability with emerging centers of international research that are going open access, and interoperability with academic repositories. The key to the future of creating this new and lasting value in the chemistry publishing web is open access to content. "We can only imagine what is possible. Actually, we have more than imagined; we have seen what is possible with organizations like HighWire. But it is certain that it will dwarf what any one company might achieve," Gold said.

[1]Kristin Antelman, "Do Open-Access Articles Have a Greater Research Impact?" *College & Research Libraries*, 372-382 (September 2004). Herbert Van de Sompel, et al., "Rethinking Scholarly Communication: Building the System That Scholars Deserve," *D-Lib Magazine* (September 2004).

7

Open Access

Participants discussed open access (OA) and OA publishing. The Budapest Open Access Initiative (BOAI)—which arose out of a meeting convened in Budapest by the Open Society Institute (OSI) on December 1-2, 2001, to accelerate progress in the international effort to make research articles in all academic fields freely available on the Internet—was discussed, and speakers talked about the OA publishing models of the Public Library of Science and the *Proceedings of the National Academy of Sciences (PNAS).*

GENERAL COMMENTS

Michael Doyle pointed out how two recent announcements put OA in the center of journal publishing. In July, a cross-party of British politicians called on the U.K. government to make all publicly funded research accessible to everyone, "free of charge on-line," he said. That same month, the U.S. House of Representatives Committee on Appropriations recommended that all NIH-funded research be made freely available six months after publication.[1]

A number of speakers commented on the American Chemical Society's policy on open access. Robert Bovenschulte explained that ACS encourages authors to link their article from their own web sites or their institutions' web sites to the article on the ACS server. He said that access to the article is free for anyone who reaches the article via this way. However, the number of free accesses per article via this path is limited to 50 during the first year, which is a total reached by hardly any articles according to Bovenschulte. One year after publication, the limit is removed.

However, finding and obtaining free articles in this way, rather than providing free access directly from the ACS publications web site is too restrictive for some. Some participants feel that ACS has been cautious about moving toward free back-files and has engendered membership resistance to it, instead of seeing it as a tool to increase membership value. A suggestion was made that ACS experiment with free back-files, and then reevaluate the matter after one year and charge accordingly. "[Open access] is a train coming down the tracks, and it ain't going to stop. React to it," Christopher Reed said. Stephen Berry added that OA is an ongoing experiment or set of experiments and it is ludicrous to think that there is a single solution. "We have to experiment," Berry said.

Michael Keller, HighWire Press, said that HighWire has a free back-issue program. Publishers in this program offer 770,000 free articles. The number increases at roughly 5,000 articles a week; all of these free articles are in science and medicine.

Martin Blume introduced the American Physical Society's policy on open access. He said that APS currently allows authors to put the PDF of an article up either on a personal or an institutional web page, if it is linked to the APS abstract. Thus, the article is essentially available free, but at the discretion of the author. According to Blume, this policy may be a first step toward immediate open access after publication, but that APS would like its journals to be totally available without access barriers. At the same time, APS currently has two journals that are open-access. One is *Physical Review Special Topics: Accelerators and Beams* that is supported by institutional sponsorship. A second one is about to start—*Special Topics: Physics Education Research*— which will be supported by author charges of about $1000 an article.

Andrea Twiss-Brooks called for self-archiving rights on

[1]More recent information on the NIH Public Access Policy is available on the Internet at *http://www.nih.gov/about/publicaccess/index.htm.*

personal and institutional web sites. She said this might be a great public relations move for publishers that allow it already, even if all scientists do not actually put their articles up on their web sites, even with permission.

The involvement of the government in open-access plans via the NIH was a red flag to some. Gordon Hammes does not think the government should be in the publication business and does not want to see NIH funds used for this purpose. There is concern about the incompleteness of the NIH database, because half of the research in many journals is not supported by NIH and would not be added to an NIH database.

Some participants questioned the six-month wait and the safety of OA to chemical literature. Peter Gregory said that being six months out of date with research, a scientist might as well not bother doing it. He questioned whether open access to chemical literature for the general public is desirable because it would result in information about explosives, propellants, pyrotechnics, bioweapons, and pharmaceuticals becoming freely available.

Most industrial librarians are really taking a "wait-and-see" approach to OA, according to Lou Ann Di Nallo. Open access might not lead to lower costs, she said. She also called for publishers to adopt a standard OA model.

Workshop participant Philip Barnett pointed out that everyone on all sides of the open access issue (publishers, researchers, librarians, and users of scientific journals) have the same goal, which is the widest possible distribution of the research literature. Yet there is often animosity among the different groups with much us versus them hostility. Complete and open communication and publicity regarding the actual cost of publishing will help reduce this animosity.

THE OPEN-ACCESS MOVEMENT

Open access and OA publishing are not one and the same, Steven Harnad explained. The genesis of the OA movement lies in June 1994, when Harnad posted what he called a "subversive proposal" to the Virginia Polytechnic Institute electronic journals mailing list (VPIEJ-L). Harnad's proposal sparked a seminal on-line debate. A part of it was later published as a book and immediately became the de facto manifesto of an embryonic OA movement.

Harnad believes that scientists, not publishers, are to blame for the fact that the community does not have OA. The intellectual content of the concept underlying open access has all the intellectual complications of a message of the following sort: "Kids, it is raining outside. Put on your raincoat. That is it. That is the intellectual content," Harnad said. That is, open access is the raincoat that will protect an author from losing research impact in the current state of restricted journal access. He said the arguments against OA—such as issues with recovering publication costs and copyright protection—are the equivalent of, "the rain is good for you. God meant us to be rained upon. Raincoats won't protect you. It is illegal to use raincoats. Raincoats will disintegrate . . . , et cetera, et cetera."

However, OA is more than free use of all scientific research, Harnad explained. The real second-order dividend of OA, above research impact, is protection from loss of research impact. For Harnad, research impact, is more than mean journal impact factor. It is closer to the dictionary sense of the word, the metaphorical meaning of impact, meaning an action that has consequences. Open access will also facilitate the essence of science, the interactive process—whether it is refereeing, commentary, or some other aspect of the collaborative, self-corrective process.

Harnad described a change in the scientist's mantra, "Publish or perish." If research is not published, it might as well be left undone, Harnad said. Research is a public collaboration, an interactive endeavor, which is why it grows and sometimes turns into benefits and applications. Yet the mantra of the scientist today is changing. It is no longer only publish or perish, but incrementally more, Harnad said. It is making research openly accessible to every would-be user on the planet who has access to the Internet.

Open-Access Publishing Versus Archiving

The BOAI defined open access and described two versions of it. Open access is toll-free, on-line, full-text access to the 2.5 million articles published each year in the approximately 24,000 peer-reviewed journals on the planet. The two varieties or two roads to open access are BOAI-1, open-access strategy number one, which is self-archiving, and was identified by Harnad as the "green road." BOAI-2, the "gold road" to OA, is to publish in journals that will provide free open access for the scientist.

Importantly, an OA journal is not one that has adopted the OA journal cost recovery model, the "author-pays" model, Harnad explained. The majority of the 1,300 open-access journals have not adopted the author-pays cost recovery model. Many of them are conventional journals that have either by principle or as an experiment made the on-line version of their content accessible toll free for all. This means that only 5 percent of the 2.5 million articles annually can be made OA. Yet, according to Harnad, it also means that the green road to open access might be a better choice right now, because the gold road might take too long. Waiting for the gold option to grow means waiting for it to create or convert and fund 23,000 more OA journals and then persuade the authors of the annual 2.5 million articles to publish in the new OA journals. This seems enormous compared to the one hurdle facing the green road: getting the authors of the 2.5 million articles to self-archive them, said Harnad.

About 92 percent of journals are green, including the "alleged bad guys like Elsevier." Harnad said he was sure that ACS would go green sooner or later, because it is already green up to 50 reprints. Generally, this means that publishers will not sue authors who self-archive, Harnad believes. For the journals that state as part of the copyright that the author may not post the article anywhere, Harnad recommends posting the preprint or the preprint and the later correction, a strategy he refers to as preprint-plus-corrigenda; it is possible to add the journal reference after the article is published.

The main purpose of OA is to maximize research impact. Research impact is a measure of the size of the research contribution, Harnad said. He said the journal impact factor is not the first or only measure of impact, or the most sensitive measure. "Impact . . . is the effect that your research has—not on the desk drawer and not on the pages of the journal in which it appears and not only on the hearts, heads, minds, and work of the researchers who are lucky enough to be at an institution that can afford the journal in which it appears—but on every potential user of that research on the planet," Harnad said.

To prove this point, Harnad and his colleague Tim Brody, at Southampton University, took 14 million articles from the ISI (Institute for Scientific Information) physics database, from 1992 to 2004, and constructed a software agent (trawler) that searched for articles contained in the physics central archive, both self-archived and non-self-archived.[2] The percentage of the overall physics literature that was found to be self-archived grew steadily from 2 percent in 1992 to 10 percent in 2001. His hypothesis proved to be right: Harnad found that self-archived articles have a higher citation impact than those behind a subscription firewall. This fact might move funding agencies and institutions toward self-archiving, he said.

Harnad explained that there is a tagging system is in place to help find articles. The 1999 Santa Fe Convention, the Open Archives Initiative (OAI) interoperability protocol, and OAI tagging all influenced the establishment of a system that ensures that every article no matter where it is, is tagged with information such as author's name and journal name, and are then drawn together into a searchable archive. According to Harnad, the self-archive cycle is as follows: An article is written and should be self-archived (but this is optional) and submitted for peer review. The peer-review cycle takes place. The final refereed version is self-archived and then the new research cycles begin.

Open access archives are not set up to be permanent, Harnad explained. OAI-compliant archives take full text in PDF, HTML, and XML format, among others. Self-archiving is not about archiving in the preservation sense of archiving, so authors do not have to worry about choosing a format for immortality. Harnad has created a search tool called ParaCite, which is being developed to locate articles from raw references using a combination of search engines including Google, OAIster, D-Lib, and others. Right now, the only way to search for the self-archive articles is with Google, Harnad said, but the overall problem is not missing search engines but rather missing content.

There is a competitive advantage in OA for research departments. A department that is completely OA would have an advantage over another department with equal quality articles, because even this little bit of an edge is enough to give it higher impact ranking.

However, the work involved in self-archiving might deter some scientists from making the effort. Christopher Reed thought that after the effort of publishing and keeping a web site, keeping an archive might be too much of an effort for academic researchers. Stevan Harnad said that the institution could take care of archiving, because it gained from the added impact.

OPEN-ACCESS PUBLISHING MODELS

Public Library of Science

Vivian Siegel described the Public Library of Science as a public charity with a mission to make the world's scientific and medical literature a freely available public resource. A way to achieve this mission is to launch open-access journals. *PLoS Biology* was launched in 2003, *PLoS Medicine* in the fall of 2004. Publication is the final and often the only tangible product of research. In an electronic era, it is possible to think about publishing as service providing and assign a fixed cost to the value that publishers add.

It is in the interest of funding agencies and institutions to ensure that the final product of research, a published manuscript, is available to everyone, Siegel said. PLoS helps cover those costs, in part through grants. The PLoS definition of open access is free and unrestricted on-line access to the research literature. PLoS has highly permissive usage licenses. Authors retain copyright, but sign a license. PLoS uses the Creative Commons attributions license. Finally, the papers are deposited in public databases.

PLoS Biology has gotten more than 1 million COUNTER compliant downloads of articles this year, about 100,000 downloads every month, and about 4 million hits each month. PLoS publishes about 20 papers every month. *PLoS Medicine* had 30,000 COUNTER compliant downloads in the first week of its existence. These numbers do not include downloads at PubMed Central. PLoS also reaches areas that do not have on-line access, Siegel added. There have been cases of the full PDF of the journal being downloaded (by someone other than the author or publisher) and sent to places where quick electronic access is not feasible.

Siegel was asked if *PLoS Chemistry* or *PLoS Physics* are currently being planned. She responded that PLoS is currently focused on biology and medicine, and that 2005 does not include a plan for a chemistry journal. "I hope that is enough of a nudge to the existing chemical journals to get

[2] S. Harnad and T. Brody, "Comparing the Impact of Open Access (OA) vs. Non-OA Articles in the Same Journals," *D-Lib* Magazine 10(6), (June 2004), accessed on the Internet at *http://www.dlib.org/dlib/june04/harnad/06harnad.html*.

their act together, before we decide that chemistry is just too slow and too unwilling to adopt these sorts of changes, and that we need to launch an alternative for the chemistry community as well," Siegel said

In response to the idea of *PLoS Physics*, Siegel added that physics is a very interesting example because the physics community has been effective at using its preprint server and physics archive. According to Siegel, the kind of sharing of information that already happens in the physical community at a different level leaves much to emulate.

Martin Blume expressed relief by this statement, to which Siegel replied, "I love the fact that [PLoS is] an organization of 25 people, and that I can say something like that and relieve Marty Blume, who has a much larger operation."

Proceedings of the National Academy of Sciences

Bridget Coughlin described the *PNAS* open-access experiment, in which authors have the option to pay $1,000 to have their papers be open access at the time of publication.[3] *PNAS* is the official journal of the U. S. National Academy of Sciences, which is a private, nonprofit, nongovernmental, and self-perpetuating society. However, *PNAS* does not receive funds from the Academy or from membership dues like other societies, nor does it give back to support Academy activities. *PNAS* has to budget to zero, utilizing two revenue streams: author charges of $1500 per article on average, and subscriptions, which range from a nominal $250 to just over $6,600. Four percent of what is published is in physical sciences; 8,500 papers are submitted a year, and 1 in 6 is accepted. The archive is free after six months.

Over a period of 10 weeks in 2003, *PNAS* surveyed 610 corresponding authors,[4] Coughlin explained. The journal asked if authors would be willing to pay a surcharge for their articles to be freely available on-line at the time of publication. Of the 210 authors that replied, 49.5 percent said yes, and 50.5 percent said no. The second question was what the maximum amount would be that they would be willing to pay. Most people, about 80 percent, were willing to pay the lowest denominator amount of $500; about 15 percent were willing to pay $1,000 per article. *PNAS* then polled its editorial board for this OA experiment. Of 110 board members, 22 percent responded; 84 percent said "try the experiment." *PNAS* also had the unanimous support of its Committee on Publications.

Starting in the spring of 2004, *PNAS* asked authors, after acceptance of their papers, if they would pay the $1,000 surcharge for their articles to be freely available on-line at the time of publication, in addition to their page charges. Only eight OA fees were waived, or 4.5 percent of all OA articles. The first article went up in May of 2004 (Figure 7.1). The last few issues (September-October 2004 time frame) were around 15 percent open access. This indicates that authors are "voting with their feet," Coughlin said. She believes the authors submitted to the journal as a sign of support for the open-access movement. The 15 percent OA research articles are multidisciplinary, with genetics and evolution leading, followed by geology and environmental sciences.

She added that subscribers now obtain access to all articles right away, so that paying the $1,000 fee was truly an extra cost. In 2005, *PNAS* is planning to adjust the open-access fee to $750 from $1,000, for those from an institution that has a site license, to reduce the burden on institution at large. *PNAS* is tracking the decay curves (Figure 7.2) of open-access articles versus non-open-access articles, to see if indeed there is traffic without a subscription block, but the numbers are still very small.

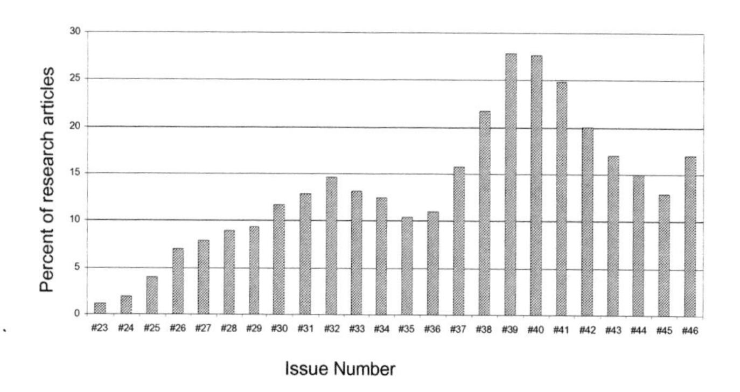

FIGURE 7.1 *PNAS* open access option uptake in 2004.

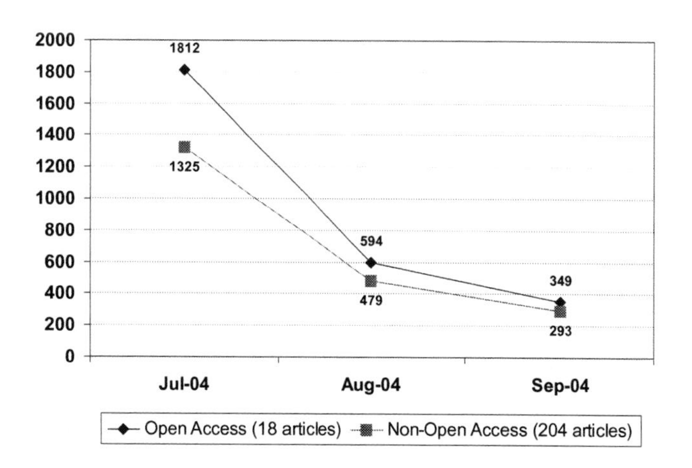

FIGURE 7.2 Average accesses per article for articles published July 2004.

[3] Nicholas R. Cozzarelli, "An Open Access Option for *PNAS*," *PNAS* 101(23):8509 (2004).

[4] Nicholas R. Cozzarelli, Kenneth R. Fulton, and Diane M. Sullenberger, "Results of a *PNAS* Author Survey on an Open Access Option for Publication," *PNAS* 101(5):1111 (2004).

Appendix A

Workshop Agenda

ARE CHEMICAL JOURNALS TOO EXPENSIVE AND INACCESSIBLE?

A Workshop Organized by the Chemical Sciences Roundtable
National Research Council

THE NATIONAL ACADEMIES
2101 Constitution Avenue, N.W.
Lecture Room
Washington, D.C.
October 25-26, 2004

MONDAY, OCTOBER 25, 2004

8:00 **Introductions and Opening Remarks,** Ned Heindel, Lehigh University

Session I Context and Overview
Ned Heindel, Chair

8:15 **Arnold Thackray,** President, Chemical Heritage Foundation

Session II What Are the Unique Scientific Journal Needs of Chemists and Chemical Engineers?
Michael Holland, Chair

9:15 **Robert Bovenschulte,** American Chemical Society

10:05 Comments and Presentations by Panel Participants

- **Christopher Reed,** University of California-Riverside
- **Patrick Jackson,** Elsevier
- **Andrea Twiss-Brooks,** University of Chicago
- **Gordon Hammes,** Duke University

11:25 Break

11:40 Discussion of Issues by Panel and Workshop Participants

12:30 Lunch

Session III Are Chemists and Chemical Engineers Receiving Needed Access to Chemical Journals?
Ned Heindel, Chair

1:35 **Ulrich Pöschl,** Technical University of Munich

2:20 Comments and Presentations by Panel Participants

- **Lou Ann Di Nallo,** Bristol-Myers Squibb
- **Michael Doyle,** University of Maryland

3:00 Break

3:15 Comments and Presentations by Panel Participants (continued)
- **R. Stephen Berry,** University of Chicago
- **Brian Simboli,**[1] Lehigh University
- **Peter Gregory,** The Royal Society of Chemistry

4:10 Discussion of Issues by Panel and Workshop Participants

5:00 Reception

Evening Presentation

The Green and Gold Roads to Maximizing Research Access and Impact: Don't Confuse Them
Michael P. Doyle, Chair

6:00 **Stevan Harnad,** University of Quebec, Montreal

TUESDAY, OCTOBER 26, 2004

Session IV What New Approaches Can Be Made to Address Chemical Sciences and Engineering Journal Needs?
Charles P. Casey, Chair

8:00 **Bridget C. Coughlin,** Proceedings of the National Academy of Sciences (PNAS)
Nicholas Cozzarelli, University of California-Berkeley, PNAS

9:05 Comments and Presentations by Panel Participants
- **Martin Apple,** Council of Science Society Presidents
- **Michael Keller,** Stanford University Press
- **Martin Blume,** American Physical Society
- **Vivian Siegel,** Public Library of Science
- **Anna Gold,** Massachusetts Institute of Technology

10:45 **Break**

Summary

11:10 Discussion of Issues by Workshop Participants

11:45 Concluding Remarks

12:00 Adjourn

[1]NOTE: Brian Simboli was unable to attend, but his presentation was made available to participants.

Appendix B

List of Participants

Prudence S. Adler, Association of Research Libraries, Washington, DC
Martin A. Apple, Council of Scientific Society Presidents, Washington, DC
Lori Barber, ScholarOne, Charlottesville, VA
Philip Barnett, City College of New York, New York, NY
Grace Baysinger, Stanford University, Stanford, CA
Jeremy Berg, National Institute of General Medical Sciences, Bethesda, MD
R. Stephen Berry, University of Chicago, Chicago, IL
Martin Blume, American Physical Society, Ridge, NY
Robert Bovenschulte, American Chemical Society, Washington, DC
Laura Brockway, Federation of American Societies for Experimental Biology, Bethesda, MD
Richard O. Buckius, National Science Foundation, Arlington, VA
Carol Carr, University of Pennsylvania, Philadelphia, PA
Charles P. Casey, University of Wisconsin, Madison, WI
Dennis Chamot, National Academies, Washington, DC
Bridget Coughlin, Proceedings of the National Academy of Sciences, Washington, DC
Nicholas R. Cozzarelli, University of California, Berkeley, Berkeley, CA
Brian Crawford, John Wiley & Sons, Hoboken, NJ (now with ACS)
Carol Cruetz, Brookhaven National Laboratory, Upton, NY
Carol Deangelo, Naval Research Laboratory, Washington, DC
Lou Ann Di Nallo, Bristol-Myers Squibb, Princeton, NJ
Michael P. Doyle, University of Maryland, College Park, MD
Arthur B. Ellis, National Science Foundation, Arlington, VA
Julie Esanu, National Academies, Washington, DC
Kenneth Fulton, Proceedings of the National Academy of Sciences, Washington, DC
Anna Gold, Massachusetts Institute of Technology, Cambridge, MA
Peter Gregory, Royal Society of Chemistry, Cambridge, UK
Elizabeth L. Grossman, U.S. House of Representatives Science Committee, Washington, DC
Gordon Hammes, Duke University Medical Center, Durham, NC
Stevan Harnad, University of Quebec, Montreal, Quebec, Canada
Victoria Harriston, National Academies, Washington, DC
Ned D. Heindel, Lehigh University, Bethlehem, PA
Steven Heller, National Institute of Standards and Technology, Gaithersburg, MD
Ahmed Hindawi, Hindawi Publishing, Cairo, Egypt
Michael J. Holland, Office of Science and Technology Policy, Washington, DC
Patrick Jackson, Elsevier, Amsterdam, Netherlands
Michael A. Keller, Stanford University Press, Stanford, CA
Lora Kutkat, National Institutes of Health, Bethesda, MD
David Martinsen, American Chemical Society, Washington, DC
Eric Massant, Reed Elsevier Inc., Washington, DC
Kari McCarron, American Association for the Advancement of Science, Washington, DC
Patrice McDermott, American Library Association, Washington, DC
Jack Morgan, Purdue University, West Lafayette, IN
Parry M. Norling, Chemical Heritage Foundation, Wilmington, DE
Gwen Owens, Georgetown University, Washington, DC
Paul Peters, Hindawi Publishing, Cairo, Egypt
Barbara Kline Pope, National Academies Press, Washington, DC
Ulrich Pöschl, Technical University of Munich, Munich, Germany

Christine R. Rasmussen, National Academies, Washington, DC
Christopher A. Reed, University of California, Riverside, CA
William S. Rees, U.S. Department of Homeland Security, Washington, DC
Sophie Rovner, Chemical & Engineering News, Washington, DC
James Schuttinga, National Institutes of Health, Bethesda, MD
Leah Solla, Cornell University, Ithaca, NY
Vivian Siegel, Public Library of Science, San Francisco, CA
Sarah Tegen, Proceedings of the National Academy of Sciences, Washington, DC
Arnold Thackray, Chemical Heritage Foundation, Philadelphia, PA
Andrea Twiss-Brooks, University of Chicago, Chicago, IL
Paul Uhlir, National Academies, Washington, DC
Song Yu, Columbia University, New York, NY

Appendix C

Biographic Sketches of Workshop Speakers

R. Stephen Berry is now James Franck Distinguished Service Professor Emeritus at the University of Chicago and is also special advisor to the director for national security at Argonne National Laboratory. He received his undergraduate and graduate education at Harvard, entering in 1948 and completing his doctorate in February 1956. He was an instructor at the University of Michigan and an assistant professor at Yale from 1960 until 1964, when he moved to the University of Chicago. He has been a member of its Chemistry Department, its James Franck Institute, the College, and the School of Public Policy Studies. His scientific activities have involved both experimental and theoretical studies. They have included studies in electronic structure of atoms and molecules, atomic and molecular collisions, chemical kinetics, chaos and regularity, atomic and molecular clusters, thermodynamics (especially of finite-time processes), and most recently, protein dynamics. His activities in areas of public policy have involved efficient use of energy and resources, science education at the middle and high school level, science and law, and of course the distribution of and access to scientific information.

Berry is married, with three children and seven grandchildren. He continues to enjoy skiing, hiking, and fly fishing, as well as music and photography.

Martin Blume is editor-in-chief of the American Physical Society, on leave from his position as senior physicist at Brookhaven National Laboratory. He received a B.A. from Princeton and a Ph.D. in theoretical solid-state physics from Harvard. At Brookhaven he has served as head of condensed matter theory, chairman of the National Synchrotron Light Source Department, and deputy director of the laboratory. He has also held a joint appointment as professor of physics at the State University of New York at Stony Brook. Since 1997 he has been editor-in-chief of the American Physical Society, with responsibility for all of the *Physical Review* journals, *Physical Review Letters*, and *Reviews of Modern Physics*. The challenge of electronic publishing and associated questions of intellectual property, archiving, peer review, cost containment and recovery, and provision of journals to all who need them are among those that must be addressed in this time of change in scholarly communication.

Charles P. Casey received his early education in St. Louis, Missouri (B.S. in chemistry, St. Louis University, 1963). His graduate research with George M. Whitesides at the Massachusetts Institute for Technology (MIT) was on organocopper compounds. After receiving his Ph.D. in 1967, he spent several months at Harvard University as a National Science Foundation (NSF) fellow in the laboratories of Paul D. Bartlett. In 1968, he joined the faculty at the University of Wisconsin-Madison where he is now Homer B. Adkins Professor of Chemistry and Steenbock Professor in the Physical Sciences. He was department chair at Wisconsin from 1998 to 2001. He was President of the American Chemical Society (ACS) in 2004.

Professor Casey's research focuses on mechanistic organometallic chemistry. The mechanisms of important catalytic processes including hydroformylation, hydrogenation, and alkene polymerization are being explored. His group has characterized d0 yttrium-alkyl-alkene complexes as models for the key intermediate in metallocene-catalyzed alkene polymerizations. Earlier work involved metal-carbene-alkene complexes and their role in both cyclopropanation and olefin metathesis, chelating diphosphines with wide natural bite angles as effective ligands for highly regioselective hydroformylations, and heterobimetallic compounds. He is author of more than 250 papers in organometallic chemistry.

Dr. Casey is a member of the National Academy of

Sciences and the American Academy of Arts and Sciences and a fellow of the American Association for the Advancement of Science. He has received the Alumni Merit Award from St. Louis University, an Alexander von Humboldt Senior Award, a fellowship from the Japan Society for the Promotion of Science, the Arthur C. Cope Scholar Award of the ACS, and the ACS Award in Organometallic Chemistry; he was a National Science Council Distinguished Lecturer in Taiwan.

Lou Ann Di Nallo is associate director, Information and Knowledge Integration, at Bristol-Myers Squibb (BMS). She directs the content integration and access function, which includes 19 people at five sites serving scientists and knowledge workers within the Pharmaceutical Research Institute and throughout the company. She leads content development and library systems in physical and virtual libraries as well as client services (document delivery, subscriptions, training, and marketing). She has been involved with electronic information over her entire career in a variety of positions in both for-profit and nonprofit settings. Prior to joining BMS she was electronic resources manager at the Hagerty Library, Drexel University, where she also taught undergraduate and graduate information courses as an adjunct faculty member.

Michael Doyle received his B.S. degree from the College of St. Thomas in St. Paul, Minnesota, and his Ph.D. degree from Iowa State University. Following a postdoctoral engagement at the University of Illinois at Chicago Circle, he joined the faculty at Hope College in Holland, Michigan, in 1968. In 1984, he moved to another undergraduate institution, Trinity University in San Antonio, Texas, as the Dr. D. R. Semmes Distinguished Professor of Chemistry, and 13 years later he went to Tucson, Arizona, as professor of chemistry at the University of Arizona and vice president of Research Corporation. He came to the University of Maryland as professor and chair of the Department of Chemistry and Biochemistry in 2003. Doyle has been the recipient of a Camille and Henry Dreyfus Teacher-Scholar Award (1973), a Chemical Manufacturers Association Catalyst Award (1982), the American Chemical Society Award for Research at Undergraduate Institutions (1988), Doctor *Honoris Causa* from the Russian Academy of Sciences (1994), Alexander von Humboldt Senior Scientist Award (1995), the James Flack Norris Award for Excellence in Undergraduate Education (1995), and the George C. Pimentel Award for Chemical Education (2002). He has written or coauthored 10 books, and 21 book chapters, and he is the coauthor of more than 250 research publications. Through his role in the creation of the Council on Undergraduate Research, the National Conferences on Undergraduate Research, and other organizations and studies, he is knowledgeable about the environment for research at predominantly undergraduate institutions.

Peter Gregory is managing director of the Royal Society of Chemistry's publishing operation, based in Cambridge, United Kingdom. Prior to this he worked for Wiley-VCH in Germany where he was responsible for the chemical engineering, industrial chemistry, and materials science programmes. He was also editor-in-chief of the journal *Advanced Materials* for more than 13 years after leaving his career in research. Peter therefore has been a researcher, an author, a referee, an editor, a commercial publisher, and now a not-for-profit publisher.

Gordon G. Hammes is the University Distinguished Service Professor of Biochemistry at Duke University. He joined the faculty at Duke in 1991 and served as vice chancellor for Medical Center academic affairs from 1991 to 1998. He was a faculty member at MIT and Cornell University prior to his appointment at Duke University. Dr. Hammes' awards and honors include an award in biological chemistry from the American Chemical Society (1967); he is a member of the National Academy of Sciences (1973), a member of the American Academy of Arts and Sciences (1974), and a National Institutes of Health (NIH) Fogarty Scholar (1975-1976); he received the 2002 William C. Rose award of the American Society for Biochemistry and Molecular Biology. He has published more than 225 scientific publications, including two books on chemical kinetics, a book on enzyme catalysis and regulation, and a book on thermodynamics and kinetics for the biological sciences. Dr. Hammes received his doctorate in 1959 from the University of Wisconsin, Madison, and was an NSF postdoctoral fellow at the Max Plank Institut, Göttingen, Germany, from 1959 to 1960. During his professional career, Dr. Hammes has been involved in various education and training programs, was president of the American Society for Biochemistry and Molecular Biology, and served on NIH training grant and research panels.

Stevan Harnad was born in Hungary, did his undergraduate work at McGill University and his graduate work at Princeton University, and is currently Canada Research Chair in Cognitive Science at the University of Quebec, Montreal. His research is on categorization, communication, and cognition. Founder and editor of *Behavioral and Brain Sciences* (a paper journal published by Cambridge University Press), *Psycoloquy* (an electronic journal sponsored by the American Psychological Association) and the CogPrints Electronic Preprint Archive in the Cognitive Sciences, he is past president of the Society for Philosophy and Psychology, and author and contributor to more than 150 publications, including *Origins and Evolution of Language and Speech* (NY Academy of Sciences, 1976), *Lateralization in the Nervous System* (Academic Press, 1977), *Peer Commentary on Peer Review: A Case Study in Scientific Quality Control* (Cambridge University Press, 1982), *Categorical Perception: The Groundwork of Cognition* (Cambridge University Press, 1987), *The Selection of Behavior: The Operant Behaviorism* *of BF Skinner: Comments and Consequences* (Cambridge University Press,1988) and *Icon, Category, Symbol: Essays on the Foundations and Fringes of Cognition* (in preparation).

Patrick Jackson is publishing director, Chemistry and Chemical Engineering, Elsevier. His background is in the natural sciences, and he has worked for more than 30 years in the scientific, technical, and medical (STM) publishing industry in various editorial and management functions in the life sciences, clinical, and chemical sciences areas. He is currently responsible for the strategic and operational development of Elsevier's primary publications in chemistry and chemical engineering, with responsibility for about 100 core chemistry journals and more than 80 new book and major reference work publications per year. He is physically located in Amsterdam, The Netherlands.

Michael A. Keller is the Ida M. Green University Librarian, director of Academic Information Resources, publisher of HighWire Press, and publisher of the Stanford University Press. These titles touch on his major professional preoccupations: commitment to support of research, teaching, and learning; effective deployment of information technology hand-in-hand with materials; and active involvement in the evolution and growth of scholarly communication. He may be best known at present for his distinctively entrepreneurial style of librarianship. As university librarian, he endeavors to champion deep collecting of traditional library materials (especially manuscripts and archival materials) concurrent with full engagement in emerging information technologies.

Keller was educated at Hamilton College (B.A. biology-music 1967), State University of New York (SUNY), Buffalo (M.A., musicology, 1970)), SUNY, Geneseo (M.L.S., 1971), and SUNY, Buffalo (all but dissertation Ph.D., Musicology). From 1973 to 1981, he served as music librarian and senior lecturer in musicology at Cornell University and then in a similar capacity at University of California (UC), Berkeley. While at Berkeley, he also taught musicology at Stanford University and began the complete revision of the definitive music research and reference materials, an annotated bibliography popularly known as Duckles in honor of its original compiler. Yale called him to the post of associate university librarian and director of collection development in 1986. In 1993, he joined the Stanford staff as the Ida M. Green Director of Libraries. In 1994, he was named to his current position of university librarian and director of academic information resources. In 1995, by establishing HighWire Press, he became its publisher, and in April 2000, he was assigned similar strategic duty for Stanford University Press.

Ulrich Pöschl is the head of the Aerosol Research Group at the Institute of Hydrochemistry, Technical University of Munich, Germany (*http://www.ch.tum.de/wasser/aerosol*). He studied chemistry at the Technical University of Graz, Austria, and worked as a postdoctoral fellow and research scientist at the Massachusetts Institute of Technology, Cambridge, Massachusetts, and at the Max Planck Institute for Chemistry, in Mainz, Germany. His current research and teaching activities are focused on the effects of aerosols on atmospheric chemistry and physics, climate, and public health (field measurements, laboratory experiments, and modeling of aerosol particle composition, structure, reactivity, and water interactions). As the initiator and chief executive editor of the open-access journal *Atmospheric Chemistry and Physics* (*ACP*, www.atmos-chem-phys.org) he started and established an innovative and successful initiative for improved scientific publishing and quality assurance in collaboration with a globally distributed network of coeditors. Moreover, he serves as the president of the Atmospheric Sciences Division of the European Geosciences Union (EGU).

Christopher A. Reed is distinguished professor at the University of California, Riverside. He obtained his Ph.D. at the University of Auckland, New Zealand, in 1971 and, after postdoctoral studies at Stanford University, served on the faculty of the University of Southern California for 25 years. His research interests span inorganic, organic, and physical chemistry. His current work involves carboranes and the synthesis of the strongest known Brønsted acids. His research has been recognized by Sloan, Dreyfus Teacher-Scholar, Guggenheim, and senior von Humboldt awards. He is a fellow of the American Association for the Advancement of Science (AAAS) and serves on the Executive Board of the ACS Division of Inorganic Chemistry and the Editorial Advisory Boards of *Chemical Communications, Accounts of Chemical Research,* and *Heteroatom Chemistry.* Essays on the chemical literature include "Drowning in a Sea of Refereed Publications" in *Chemical and Engineering News,* (January 29, 2001); "Electronic Access to Journals" in *Chemical and Engineering News,* (October 29, 2002); and "Publish *and* Perish" in the *Chronicle of Higher Education* (February 20, 2004). These can be accessed at *http://reedgroup.ucr.edu.*

Brian Simboli is a science librarian at Lehigh University and a part-time writer. He attended Swarthmore College for his undergraduate work and received a Ph.D. in philosophy from Notre Dame, as well as an M.S. in library science from Drexel University.

Arnold Thackray received his Ph.D. from Cambridge University. He has held faculty appointments in Oxford (visiting fellow, All Souls College), Cambridge (fellow, Churchill College), and at the London School of Economics, Harvard, the Institute for Advanced Study, the Center for Advanced Study in the Behavioral Sciences (Stanford), and the Hebrew University of Jerusalem. He was founding chairman of, and

Joseph Priestley Professor in, the Department of History and Sociology of Science at the University of Pennsylvania.

Thackray's scholarly interests lie in the historiography of science and in understanding technology, medicine, and science as elements of modern culture. He served as editor of *Isis*, the official journal of the History of Science Society, for seven years, and as editor of the society's newer journal, *Osiris*, for ten years. He has been active in the public life of scholarship, serving on a number of boards, including that of the American Council on Education, and is a former president of the Society for Social Studies of Science and was treasurer of the American Council of Learned Societies for more than a decade. Thackray is a fellow of the Royal Society of Chemistry and the Royal Historical Society, and a member of the American Academy of Arts and Sciences. He was founding director, and now serves as president, of the Chemical Heritage Foundation.

Andrea Twiss-Brooks is the bibliographer for chemistry, physics, geophysical sciences, and technology at the University of Chicago's John Crerar Library. She is active in the American Chemical Society's Division of Chemical Information, where she has served in a variety of roles, including that of division chair. Andrea has also been involved in the organization of technical symposia at ACS national meetings on topics related to chemistry publishing, including most recently a session on open-access issues in scholarly publishing. She is current chair of the American Chemical Society Joint Board-Council Committee on Chemical Abstracts Service and also a member of library advisory groups to ACS Publications and to Chemical Abstracts Service.

Appendix D

Origin of and Information on the Chemical Sciences Roundtable

In April 1994, the American Chemical Society (ACS) held an Interactive Presidential Colloquium entitled "Shaping the Future: The Chemical Research Environment in the Next Century."[1] The report from this colloquium identified several objectives, including the need to ensure communication on key issues among government, industry, and university representatives. The rapidly changing environment in the United States for science and technology has created a number of stresses on the chemical enterprise. The stresses are particularly important with regard to the chemical industry, which is a major segment of U.S. industry; makes a strong, positive contribution to the U.S. balance of trade; and provides major employment opportunities for a technical work force. A neutral and credible forum for communication among all segments of the enterprise could enhance the future well-being of chemical science and technology.

After the report was issued, a formal request for such a roundtable activity was transmitted to Dr. Bruce M. Alberts, chairman of the National Research Council (NRC), by the Federal Interagency Chemistry Representatives, an informal organization of representatives from the various federal agencies that support chemical research. As part of the NRC, the Board on Chemical Sciences and Technology (BCST) can provide an intellectual focus on issues and fundamentals of science and technology across the broad fields of chemistry and chemical engineering. In the winter of 1996, Dr. Alberts asked BCST to establish the Chemical Sciences Roundtable to provide a mechanism for initiating and maintaining the dialogue envisioned in the ACS report.

The mission of the Chemical Sciences Roundtable is to provide a science-oriented, apolitical forum to enhance understanding of the critical issues in chemical science and technology affecting the government, industrial, and academic sectors. To support this mission, the Chemical Sciences Roundtable will do the following:

- Identify topics of importance to the chemical science and technology community by holding periodic discussions and presentations, and gathering input from the broadest possible set of constituencies involved in chemical science and technology.
- Organize workshops and symposia and publish summaries on topics important to the continuing health and advancement of chemical science and technology.
- Disseminate information and knowledge gained in the workshops and reports to the chemical science and technology community through discussions with, presentations to, and engagement of other forums and organizations.
- Bring topics deserving further, in-depth study to the attention of the NRC's Board on Chemical Sciences and Technology. The roundtable itself will not attempt to resolve the issues and problems that it identifies—it will make no recommendations, or provide any specific guidance. Rather, the goal of the roundtable is to ensure a full and meaningful discussion of the identified topics so that the participants in the workshops and the community as a whole can determine the best courses of action.

[1] *Shaping the Future: The Chemical Research Environment in the Next Century*, American Chemical Society report from the Interactive Presidential Colloquium, Washington, DC, April 7-9, 1994.